D0327421

Science and the Quest for Meaning

Science and the Quest for Meaning

Alfred I. Tauber

BAYLOR UNIVERSITY PRESS

Waco, Texas 76798

Cover Design by Steve Scholl, The WaterStone Agency
Cover Images: Leonardo DaVinci's Vitruvian Man ©istockphoto.com/jodiecoston; Ancient map on leather ©istockphoto.com/muratsen. Used with permission.

Library of Congress Cataloging-in-Publication Data

Tauber, Alfred I.
Science and the quest for meaning / Alfred I. Tauber.
p. cm.
Includes bibliographical references and index.
ISBN 978-1-60258-210-1 (hardcover : alk. paper)
1. Science--Philosophy. 2. Science and the humanities. I. Title.
Q175.T2245 2009
501--dc22
2009010375

Printed in the United States of America on acid-free paper with a mimimum of 30% pcw content.

for Hal

Science cannot decide questions of value, that is because they cannot be intellectually decided at all, and lie outside the realm of truth and falsehood.

Bertrand Russell, *Science and Religion*

I do not see reality as morally indifferent: reality, as Dewey saw, makes demands on us. Values may be created by human beings and human cultures, but I see them as made in response to demands that we do not create. It is reality that determines whether our responses are adequate or inadequate.

Hilary Putnam, *Jewish Philosophy as a Way of Life*

The true man of science will have a rare Indian wisdom—and will know nature better by his finer organization. He will smell, taste, see, hear, feel, better than other men. His will be a deeper and finer experience. We do not learn by inference and deduction, and the application of mathematics to philosophy but by direct intercourse. It is with science as with ethics—we cannot know truth by method and contrivance—the Baconian is as false as any other method. The most scientific should be the healthiest man.

Henry David Thoreau, *Journal*, October 11, 1840

Contents

Acknowledgments

This book joins the two sides of the divide between science and the humanities. I have arrived at my own perspective from two vocations. As a research physician, I spent twenty years of my career engaged in investigations of the biochemistry of the inflammatory response. That work spanned the fields of free-radical chemistry, protein chemistry, cell biology, and molecular biology. With some overlap, I devoted another twenty years to the history and philosophy of nineteenth- and twentieth-century biology, with particular studies devoted to the development of immunology's basic theories, the place of reductionism in biomedicine, and most recently, the intersection of science and moral philosophy. *Science and the Quest for Meaning* in many ways completes the project that began with my transition from a laboratory scientist to a philosopher and historian of science. Building a bridge between these two domains has been a richly rewarded intellectual endeavor, one assisted by many students, colleagues, and friends. To acknowledge all those who have aided me simply resides beyond my sorry memory, but Bob Cohen, Hilary Putnam, and John Stachel deserve special acknowledgment for their support and guidance.

I have developed my interpretation of science as an intellectual and cultural activity from a privileged position. Since 1993, I have had the good fortune of surveying science studies as Director of Boston University's Center for Philosophy and History of Science. The Center, created in 1960, is the oldest and one of the most distinguished forums for scholarly discussions of the history, philosophy, and sociology of science, mathematics, and logic. This book, in large part, is a distillation of my intellectual excitement stimulated by the erudition and creative interpretative insights of our visiting scholars and resident fellows. The Center's Boston Colloquium for Philosophy of Science has hosted over 800 lectures since 1993! In the rich dialogue that these presentations have stimulated, I have witnessed the growth and evolution of those disciplines devoted to the study of science. This book is not a précis or overview of science studies, but rather a report of my own views, which may fairly be judged as idiosyncratic when measured against the vast majority of studies published in this field. Indeed, I freely admit that my humanistic orientation is at odds with the temper of the times. All the more reason to offer this analysis!

More immediately, this book crystallized around two of my lectures and the discussions surrounding them. The first, "Science and the Humanities," was delivered at the European Regional Conference of the Board of Governors of Tel Aviv University (Berlin, April 1, 2005). That paper was expanded into a second presentation, "Science and Reason, Reason and Faith: A Kantian Perspective," for the Herbert H. Reynolds Lectureship at Baylor University (January 31, 2006) and later published (Tauber 2007). Appropriate to its genesis, this book has been nurtured by Carey Newman and the staff of Baylor University Press, and I thank them for their professionalism and effective efforts on behalf of this project. I have also profited from my graduate and undergraduate seminars based on this text at Boston University (2005) and Tel Aviv University (2006), where my students helped me frame and articulate the issues discussed in the book. I am particularly appreciative to the skeptics who challenged my thesis and provoked me to more clearly argue its case. I will not enumerate the many colleagues and friends with whom I have profitably explored the specific issues presented in this book, but I extend a special thanks to those who read this manuscript at various stages of its production: Chalmers Clark, Menachem Fisch, Scott Gilbert, Charles Griswold, Walter Hopp, Adi Ophir, Jonathan Price, Peter Schwartz, Steve Scully, and John Stachel. I must single out Roger Smith, who, for many years, has prodded me to rethink my positions and to probe more deeply into the issues described here and in related publications. Those generous efforts are much appreciated. Finally, I thank my wife, Paula

Fredriksen, who has always been my most consistent and effective interlocutor. This study is dedicated to my friend Hal Churchill, in the hopes he will read it. Hal personifies the caring physician, a man devoted to science and its humane application, so I regard the message conveyed here a testament to his own commitments.

Portions of this book have been culled from previously published material where I dealt with various areas of contemporary science studies. Some issues, such as the aesthetic dimension of scientific inquiry (Tauber 1996a) and the "bio-political" character of molecular biology and certain aspects of its application are long-standing interests of mine (Sarkar and Tauber 1991; Tauber and Sarkar 1992, 1993; Tauber 1999c). The more specific project of understanding the fact/value relationships operative in nineteenth-century science appeared in my examination of Henry David Thoreau's response to the ascendant positivism of that era (2001). I extended that case study in *Patient Autonomy and the Ethics of Responsibility* (2005a), in which I considered the relationship of medicine's epistemology and ethics. Key passages from various works have been excerpted and reformulated for this book. Accordingly, I hereby acknowledge permission to reprint edited portions of these previous publications: "Science and reason, reason and faith: A Kantian perspective" (2007, 307–36); "Ecology and the claims for a science-based ethics" (1998, 185–206); review of *The Historiography of Contemporary Science and Technology* (1999b, 384–401).

Boscawen, New Hampshire
November 2008

Introduction

Concerning Scientific Reason

If science is not to degenerate into a medley of ad hoc hypotheses, it must become philosophical and must enter upon a thorough criticism of its own foundations.

Alfred North Whitehead, *Science and the Modern World*

Growing up in the Sputnik era during the 1950s, I enjoyed what appears now to have been a unique education. Science assumed an importance hitherto unimagined prior to the Soviet challenge, and to prepare the country for possible assault, beside air raid simulations, I studied "new math" and was enrolled in advanced science courses. Drilled in facts, disciplined in scientific method, and buoyed by the wonder of nature, I saw a future bright with the scientific enterprise. Perhaps I too would become an investigator. In that spirit, an even more important foundation was being set for myself, namely a sense that science offered something close to true knowledge as the technical mastery of nature proceeded with breath-taking achievements. Weren't we about to embark for the moon? Such mammoth enterprises were undertaken under the banner of truth, and truth was attained through objective methods. It seemed that science defined its own domain, and not only remained insulated from common human foibles, but followed methods that revealed Truth. This "Legend" (Kitcher 1993), simple and distorted as it might be, nevertheless was cherished by its believers. Indeed, every Saturday morning Mr. Wizard appeared on television to elucidate nature's mysteries, and thereby confirm the precepts taught to me. The shades of grey were apparent on the screen; the colors were not. That was the world in which I awakened, one seemingly simpler than today.

Of course, doctrine is fated for refutation, which already had commenced even as I was learning the solar system model of the atom. The philosophy of science that framed my generation's education still promoted a stark nineteenth-century positivism. The term positivism refers to a philosophy of "positive" (objective) knowledge, which means, simply, that valid knowledge is scientific; facts are the currency of knowledge; accordingly, forms of knowledge that do not subscribe to the scientific method cannot be validated. Positivism thus rested, ultimately, on the separation of "facts" from "values."

Values were usually considered a catchall for subjectivity, but of course, *epistemic* values—those values that made facts, facts (e.g., objectivity, neutrality, coherence, parsimony, predictability)—were integral to the scientific enterprise.[1] And beyond recognizing the diverse values that must be employed to create objective facts, the overlap of so-called "subjective" values in constructing scientific knowledge has increasingly become apparent.

Indeed, much of the scholarship over the past fifty years characterizing scientific practice and theory formation has shown that the relationships between facts and values, even within the narrow confines of laboratory investigations, cannot be neatly divided between "objective" and "subjective" domains. And when the doors of the laboratory are flung open and the applications of research are considered, the complex relationship of facts and values becomes even more convoluted. Factoring out the ever-present commercial aspects of investigations, as well as the various agendas of government-supported research for military or economic gain, the objective/subjective schema simply defies the social and conceptual realities of scientific inquiry.

The irony of science portraying itself as a fantasy—a restful space for logic and rational deliberation as sole determinants of research, one that would achieve some utopian respite from the tribulations of human-derived confusion—is a story which has been told from many points of view. Here I will narrate how the conceptual scaffolding supporting the castle in the sky fell and then offer a summary of the post-Sputnik description that replaced it. Coupled to that dismantling of the Legend, we will survey the cultural war that commenced with the reports of revisionist historians, sociologists, and philosophers of science. Citizen activists joined them under the belief that characterizing (and controlling) science was too important to leave to the self-appraisals (and choices) of scientists alone. This book is about that seismic intellectual and political shift, and perhaps, in a sense, it is a revised narrative about my own youthful naiveté.

Philosophically, the positivist program began to crumble during the early 1950s (Friedman 1999), and with the loss of its intellectual dominance, a critical chorus challenged the authority of a doctrinaire scientific method and its hegemonic form of knowledge. From that dissenting position, science appeared to have spun into its own orbit. Instead of celebrating the polyphonic contributions of all sectors of scholarship, competing science/anti-science camps assembled along academic lines, in which the scientific illiteracy of the literati and the deafness of scientific technocrats precluded meaningful dialogue. C. P. Snow famously described this rift in terms of "Two Cultures," inasmuch as the sociologies and modes of discourse of each group had radically diverged (Snow 1959/1964). A more damning appraisal remained unwritten: because of its success and its independence of the larger philosophical context from which it emerged, science was regarded as an unruly adolescent, full of itself, brimming with confidence and even arrogance, overflowing with its power and promise.

Having assumed a unique place in the academic pantheon, science pursued its own agenda with confidence and little concern for relating to its "distant relations." This division was well underway by the mid-nineteenth century, when the twin forces of professionalization and positivism drove the scientist to distant lands, where he learned new languages, adopted peculiar mores, and cultivated particular industries. As Wilfrid Sellars noted (writing as a philosopher):

> The scientific picture of the world *replaces* the common-sense picture . . . the scientific account of "what there is" *supersedes* the descriptive ontology of everyday life. . . . [I]n the dimension of describing and explaining the world, science is the measure of all things, of what is that it is, and of what is not that it is not. (Sellers 1997, 82–83; emphasis in original)

Here, "common-sense" is a placeholder for all those modes of knowing eclipsed by the triumph of science's worldview.

Humanists feared an imbalance in two domains. The first was intellectual: Humanists viewed science as assuming imperialistic ambitions in the attempts to apply its methods and logic in arenas where heretofore it had not ventured. This so-called scientism (the belief that virtually anything worth knowing or understanding may be approached scientifically and given scientific explanation) had been on the positivists' agenda for over a century, but by mid-twentieth century, humanists actively charged that such scientistic claims were by their very nature fallible, since radical objectivity had repeatedly been deflated by showing how pernicious cultural determinants influenced scientific inquiry and interpretation. Despite its failures, the positivism that dominated the natural and social sciences asserted a rigid factuality to what constituted knowledge, and that standard, broadly applied, would devalue other forms of inquiry. Thus, as a purely intellectual conflict, most scientists and humanists found themselves on different sides of the demarcation lines outlined by the positivist program.

The political and social domain was the second area where science posed a threat to the humanities. Despite the technical achievements of science, humanists rightly feared the imbalanced influence of the science "lobby," whose authority rested on the economic bounty indebted to scientific advances. The Two Culture divide was, consequently, also an expression of how science, largely as a result of its material successes, increasingly dominated public policy decisions and educational resources. The social apparatus that supported the scientific enterprise ranged from the educational reform stimulated by the Sputnik challenge to scientific industries promoting their vested interests. Beyond the technology sold domestically in the West, these industries were prominently energized by what Eisenhower menacingly described as a military-industrial complex, which prominently displayed its products in Vietnam and later in Iraq. Many were troubled by the danger of misplaced applications (like nuclear power) and, even more, by a kind of political arrogance that seemed to accompany the power of unbridled technology. These matters, while germane, are not our subject. Here, suffice it to note that by the end of the 1950s, science education dominated other forms of knowing, so that a gentle species of scientism seeped into the schools educating the Baby Boomers. Dissenting voices, of course, attempted to find the humane within the scientific enterprise (Conant 1953; Bronowski 1956), but with nuclear war threatening civilization on the one hand, and the recent conquest of polio on the other, science (albeit, a particular positivist vision of it) and derivative technologies were grabbing all the headlines—and the money.

As science assumed a new degree of independence based on its ever-increasing authority, the disciplines of history and philosophy of science morphed into a new species. They filled a gaping hole. After all, as Thomas Kuhn noted, scientists generally are not interested in their own histories, much less the philosophy undergirding their discipline (Kuhn 1962, 1970). But beyond this professional separation, the respective ways of thinking seemed foreign to each other, thus cross-fertilization had become increasingly barren. Ironically then, coincident with Snow's critique, the original cultural divide began to mend in an unpredictable way as interdisciplinary studies of science achieved new sophistication. Philosophers, historians, and sociologists of science pursued an ambitious program to characterize the laboratory as an intellectual and cultural activity devoid of positivist conceits. No longer was science allowed to perform insulated from outside scrutiny. Consequently, the Two Cultures mentality that Snow and others had so recently identified quickly collapsed as critics of science asserted challenging interpretations of what scientists did, what philosophical structures they employed, and how they conducted themselves. Today, much of what serves as debate about what science is and what it does may be reduced to those who seek to demarcate the various kinds of truth claims arising from different intellectual cultures from those who seek to bridge the apparent chasm between them.

Indeed, science was wrenched back from its isolated status, and the Two Cultures were melded back to one with a vengeance. Paul Feyerabend in *Against Method* (1975) attacked the sacrosanct status of scientific rationality; Kuhn's *Structure of Scientific Revolutions* (1962) rejected claims to orderly scientific progress; and Michael Polanyi's *Personal Knowledge* (1962) offered a more comprehensive appreciation of scientific thinking than that proffered by positivist philosophies of science. These works marked the beginning of a new movement to study science in a broadened humanistic and sociological context, which employed analytic tools quite alien to the then current "internal" approaches that followed the positivist line without dissent.

While the boundary between science and nonscience served as a critical nexus of positivist thought, the post-Kuhnian critique opened a schism for all to see. Indeed, the self-confident posture of the "scientific" suffered from these radical criticisms, and although the work of Kuhn, Feyerabend, and Polanyi took a generation to take hold, their cardinal lessons have gained legitimacy in hard fought debates. After all, scientific knowledge has increasingly defined natural realities, and, in the process, such knowledge, putatively objective and neutral, had assumed secularism's closest approximation to truth. Facts are sacred. Debate might ensue as to what those facts mean and how we might apply them to social and economic policy, but such circumspection only highlights how interpretation, undergirded by a vast array of values, determines the use of knowledge in the *political* realm. That insight and the caution it has generated, date, at least, to the early part of the twentieth century. However, something new emerged during the twentieth century's closing decades.

Beyond active public debate over the direction of scientific inquiry, the new critics challenged doctrine—the very notion that one could defend a *method* of scientific inquiry built on a firm demarcation of facts and values. No longer were *facts* simply facts. What might appear as a fact in one context might be revealed as only a factoid, or perhaps not even a reliable claim or report. But a more fundamental issue appeared in discussions about objectivity and the character of facts: facts are *always* processed—interpreted, placed into some over-arching context—whether in debate about a scientific theory or an argument about social policy. Inextricable from context, facts must assume their meaning from a universe of other valued facts. When applied to scientific methods and the logic that governs investigation, the traditional orthodox method, which promised hard, neutral facts *and* derivative truths, revealed an Emperor disrobed. In that scandal, the credibility of scientific testimony became contentious. Scientific facts, which hitherto had been thought of as neutral, were now recognized as taking on different meanings depending on factors far removed from a narrow construal of their placement in a model or theory. Thus the neutrality of the facts that framed the issue in question became suspect, and so the fight over the significance of scientific findings took place not only in traditionally prescribed professional circles, but in the courts and legislatures as experts pitched themselves in service to one socio-political position or another.

Despite the reiterated disavowals of a value-laden science, critics have exposed the neutrality of science as a useful conceit. Increasingly, citizens are maintaining a vigilant watch over science's aspirations and successes. Such activists no longer accept as dogma the claims and promises of a growing scientific lobby. For instance, in 1993, critics successfully halted the superconducting collider project, the exemplar of Big Science, in what some regarded as antiscientific conservatism (threatening United States leadership in elementary particle physics), and others saw as appropriate constraint of a ravenously imperialistic venture. This debate seems to have generated a different kind of activism than that of the previous attacks on what had been perceived as unbridled technology (e.g., nuclear power or environmental pollutants). The distinction between science and its product, technology, traditionally afforded scientists the space to pursue their research in the interests of advancing knowledge, leaving its application to another public forum. However, in recent decades the notion of science in the pursuit of knowledge, research for the pursuit of truth, has been challenged by a ravenous technology that has come to dominate science, reversing the historical relationship between basic investigation and its application (Forman 2007). On this view, science no longer enjoys such latitude in its enterprise and can no longer be regarded as some colony of its motherland, protected from intruders that might invade its sanctioned ways and profitably tap its resources. Indeed, a contemporary portrait of science must account for its social character in a complex calculus, where science is understood as subject to powerful economic and political interests, and, in turn, pursues its own agenda for its own particular gain. This understanding rests on a multi-layered cultural and intellectual history.

How science is understood determines how its knowledge is applied. If the positivist program asserts that such and such is the case, and if the public relies on the certainty of such claims, then the authority of scientific knowledge achieves high standing. If, on the other hand, scientific knowledge is regarded as always fallible and its methods always in question, then scientific claims will be viewed with more circumspection. Following Kuhn and Feyerabend, a deluge of sociologically oriented critics looking at what scientists actually did, as opposed to the idealized philosophical claims made on their behalf, brought science well within the fold of other forms of truth seeking.[2] Science-studies philosophers, historians, and sociologists converged on depicting scientific practice on a pragmatically based epistemology (replacing reified method and verification). Consequently, the insularity of the laboratory and the truth claims made under its mantle were increasingly called into question. These reassessments rallied "defenders" of science to protect the perceived assaults on objectivity and rational discourse made by "postmodern enemies of the Enlightenment." The resulting "Science Wars" of the 1990s brought to climax conflicts that had simmered for decades.

The intellectual studies of science complemented social forces that opened science, as an institution, to new kinds of scrutiny. As a result, science lost its privileged state: its epistemological authority lacks its former sacrosanct status; its practice no longer commands the awe enjoyed before World War II; and the queries of concerned citizens can no longer be dismissed as the ramblings of the naive. The political chorus has become more brazen in its effrontery, stopping certain kinds of science and directing others through more rigorous administrative control. A unified culture, where scientists and activists meet on equal footing, has developed in the world of public policy, but not necessarily with salutary results for utopian-minded science enthusiasts, who sought a more complete independence and authority. The Science Wars may have formally been declared in the 1990s, but the skirmishes had been passionately fought since the end of World War II. And with the recent radical politicization of science, from the global warming "debate" to public financing of stem cell research, the battles over science have achieved a feverish pitch. I choose not to argue over the specifics of these polemics, but rather to highlight how the doors of the laboratory were flung open and ponder some of the consequences of science reassessed.

General Themes

This book portrays science from a humanistic perspective, which views science not simply as an establishment seeking objective knowledge, but as a participant in the subjective interpretations of human understanding of the nature, society, and the human mind. On this view, characterizing science in a positivist modality narrowly distorts how it functions in advanced industrial societies; furthermore, such strictures misconstrue what science is. Indeed, the history of science as an idea requires both a description of an intellectual and technical enterprise, as well as an account of how the picture of reality that science presents impacts on our personal understanding of the world. That world is not just the natural environment, the human body, and the stellar sky, but includes more intimately our placement in that cosmos and the characterization of our own human nature. Most studies of science focus on science itself; here, we will examine the broader concerns to show how our current understandings of scientific knowledge impact on our deepest notions of reality.

This analysis seeks to lessen the tension between two ways of experiencing nature. On one view, science presents an objective picture, one that it has been obtained by a stark separation of subject and object. The contrasting vision, the one characterized by contemporary science studies and indebted to the romantics, understands science as melding various ways of knowing and drawing from many reservoirs of cultural influences. Each perspective accepts that science offers a unique way of depicting reality, but the former admits no subjective elements into its process, while the latter argues that the subjective remains constitutive to the scientific endeavor at every level. The difference in how we understand science, both in its process of generating knowledge and utilizing that information, is analogous to the contrast between the black and white television screen of the early 1950s, and the brilliant color and stereo sound system of contemporary high definition television. The first represents the conception of science based on a simple objectivity limited by a relatively primitive circuitry and vacuum tubes, in contrast to today's more complex understanding of science constructed from the electronics and satellite transmissions of a digital age. Early televisions have nostalgic value and historical interest, but they could not meet our desires for the most sophisticated and accurate transmission of the events we watch. Indeed, why would we cling to an outmoded technology? The implicit question posed here falls into the same pattern: Why would we accept a description of contemporary science that gives a limited, even distorted, image? The new picture we have at our disposal results from fifty years of science studies that has developed ways of understanding the complex institution we call science in ways that are radically different from the descriptions offered prior to 1960. This book characterizes those differences and draws a set of conclusions that presents science as a vibrant, personal aspect of our lives.

Science and the Quest for Meaning underscores post-positivist insights and explores how objective knowledge becomes integrated into our personal worldviews. This is an old problem. Before the rise of nineteenth-century positivism, the romantics sought to cohere the world—the subjective and objective; the positivists would disjoint that effort. I believe we must revisit the problem and seek new responses. So beyond the diverse material and social roles of science in contemporary Western societies, I am concerned here with the profound ways in which science frames the way in which we conceive the world and ourselves in it. Indeed, I maintain that the deepest conceptions of Western human being take root in the complex mulch of scientific fact and theory. The metaphysics of selfhood, society, and nature each take hold in the reality science presents to us, so on this view, the translation of knowledge to meaning represents the final step of the scientific endeavor. After all, science began with the desire to master nature *coupled* to probing the wonder of nature's mysteries for human understanding. This latter metaphysical pursuit often remains obscured by the technical triumphs of modern science. I wish to remind that this dual agenda has always guided the scientific enterprise.

To begin, we will review the diverse roles of science in contemporary Western societies and the many kinds of reason it employs. The reasons *of* science refer to a conceptual précis of scientific methods and the philosophies that explain knowledge acquisition. The reasons *for* science point to several facets: science's general (and most obvious) task of promoting technological growth; its putative neutral role in adjudicating the socio-political debates over social policy; and its pervasive impact on existential and metaphysical formulations of the character of human nature, the place of humans in nature, and the nature of being. So to capture its philosophical character, science must be regarded from at least three vantages—epistemological, ethical, and metaphysical.

Reason's configuration in science has always been contested, fitful, and confusing, and, needless to say, it has hardly achieved stability or even clear demarcations. We too must establish our own understanding of reason, and at this time, a pluralistic picture dominates. That evaluation rests on several descriptions that have emerged from recent science studies: (1) the "boundary question" depicting the elasticity of the borders defining science; (2) the constructivist elements in the production of scientific knowledge; (3) the juxtaposition of epistemic and nonepistemic values in the creation of facts and the exercise of scientific judgment; and (4) the more general changing relationships of facts and values, which interplay in the creation of scientific knowledge and its application. Each of these issues may be traced to a relaxation, if not collapse, of the fact/value distinction, which in my view clearly demarcates the positivist view of science from everything that followed positivism's demise. This book explores both how this different understanding of scientific reason emerged and the ways in which such a description accounts for science's epistemology and its socio-political applications. Indeed, discerning and then establishing science's *philosophical* position on the coordinates of fact and value links the wide expanse of contemporary science studies.

Reason has many forms and expressions, which assume their characteristic use in various contexts (i.e., scientific inquiry, discourse, and interpretation). For instance, one format considers the relationship of "reasons" and "causes" in the sense that Donald Davidson argued, namely, that reason not only explains actions, but also causes them (1980). In another context, reason functions in diverse activities, modes of thought and cognition, which address (1) the aims of science, (2) the methods employed to achieve those goals, and (3) the theories and claims arising from this venture (Laudan 1984a). Distinctive rules and logic govern these different forms of scientific rationality. Thus scientific rationality holds no single approach nor possesses an encompassing logic. A different role for reason, and, consequently, a different characterization, results from science's cultural role. Driven by diverse forces—social, political, economic, historical, cultural, and so forth—and governed by a complex range of moral, logical, aesthetic, and psychological modes of thought, science responds to many agendas. Indeed, the requirement to satisfy so many masters has stymied formal attempts to define science's reason beyond the loosest of definitions.

This complexity reflects the conceptual heterogeneity of science itself and its commanding presence in contemporary industrial societies. Science (at least as usually conceived) powerfully shapes cognition, and in one sense, defines what *is* as its methods prominently establish standards of knowledge. This authority creates a tension: on the one hand, science instantiates a particular form of Western reason (irrespective of its various modalities), which must take its place at the table with other forms of knowledge. So the question arises, How is scientific knowledge integrated into the social and psychological lives of us all? This issue—how coherence may be sought among competing individual needs, social demands, and various forms of experience—returns to the dilemma of reason's unity. Obviously different faculties of knowing are at play. The challenge is to give proper balance to each. Three key points of integration are discussed here:

(1) *The Two Cultures.* Science no longer resides outside the humanities as some distant colony of academic inquiry. Sociologists have

incorporated the laboratory within a more general sociology of knowledge; historians have shown the jolted evolution of science as anything but strictly rational in its progress; and philosophers have discounted formalism and particular logics as so much conceit, thus laying the foundation for understanding science's epistemology with a pragmatic eye. Collectively, science studies have thus demonstrated the diversity of cognitive methods and extracurricular influences governing scientific practice. These characterizations resonate with the general principles of sociology of knowledge. On this view, science has been dethroned from its special positivist pedestal, and a One Culture mentality has emerged to challenge the Two Culture picture of science in society. This new view presents science as open to the same general analysis applied to other sociologies of knowledge. However, the older picture of science pursuing its own esoteric agenda, leaves the Two Culture divide dominating *popular* conceptions of science. The perspective adopted here argues the case for One Culture. After all, while the Two Culture division framed the debate about science in the nineteenth century and well into the twentieth, now we face the challenge of understanding the myriad connections that place science firmly in its supporting culture—intellectual, political, and social.

(2) *Science as politics.* The political character of science has been exposed in a multitude of public arenas, and thus the relationship of science as a social institution with its supporting culture demands citizen control of the scientific product. Broadly construed, this latter matter is structured by the ethics of research, the political uses of scientific knowledge as applied in public policy, and the interpretation of scientific findings as understood in the context of complex human needs.

(3) *Personal knowledge.* The tensions so evident in the social sphere operate as well at the level of the individual. Indeed, the most important of my broader concerns considers various responses to making the worldview science offers one's *own*. In Descartes' formulation of *res cogitans* (mind) and *res extensa* (matter), humans self-consciously peer *at* the world—fundamentally separate and distinct. He thus framed the basic question of the modern era, namely, how might integration be achieved? This challenge—the

> ultimate question posed by modernity's preoccupation with placing oneself in an alien world—is no less than the problem of finding meaning in a world devoid of enchantment.

To describe these three domains, my project employs a distinctive architecture. Consider a room in the mansion of the mind, where reason resides divided between its diverse services to science and nonscience. This chamber we will call *Reason Divided.* On this commonly conceived division, science employs one kind of objective reason—strict and rigorously defined by logical rules—and other forms of inquiry must rely on different forms of reason that address the subjective.[3] Admittedly this stark contrast only holds at the extremes. After all, legal argument and journalism, for example, model themselves on the same kind of objectivity one hopes to find in the laboratory. Thus scientific thinking frequently serves as the standard by which other disciplines aspire to be rational. So note, the rationality and logic employed by scientists is not at issue. Instead, critics have largely focused their attacks on "scientism," which would inappropriately apply a scientific orientation to subjects not accessible to such methods.

Scientism expresses itself in at least three ways: (1) the sciences better capture "reality" than other disciplines (e.g., for E. O. Wilson, science will eventually not only unify thought but reduce human behavior and culture to a biological formulation [1998]); (2) scientific methodology trumps other forms of knowing (e.g., Ernest Rutherford's quip, "there is physics and there is stamp collecting" [Blackett 1963, 108]); and (3) *if* philosophical problems essentially reduce to scientific problems, then "philosophy of science is philosophy enough" (Quine 1953d/1976, 149). Each of these assertions rests on the fundamental notion that reality is consonant with, if not superimposable on, the picture science offers.

If contemporary science studies have a general ethos, I think it is to refute these presumptions, a task largely accomplished. Indeed, we are at the end of positivism's fall (Zammito 2004). Not only have the domains of science assumed new, fuzzy, and mobile borders, the very judgment of scientific reason has been opened to include logics constructed from various sources and directed towards diverse ends. In short, science has many reasons, some of which are strictly confined to rules construed as objective and others not. Here, pragmatism rules. So in the adjoining room, *Reason Unified,* science is part of a greater domain of inquiry. This side seeks a science fully integrated with the larger humanistic inquiry, incorporating personal and social values instead of excluding them, seeking synthesis in place of insular division, and employing eclectic modes of knowing instead of restricted and narrow means of knowledge acquisition. A group of values for this venture command attention quite different from those advocated by the positivists and the most recent "defenders of science" who still hold to some form of neopositivism.

Contemporary science studies provides the hinge of my project that swings open the door from the first room to the second. As already mentioned, instead of a Two Cultures mentality, I prefer to explore the possibility of finding a common space for two communities that have too long remained alienated from each other. Two expansive views appear as the portal opens: First, scientific practice follows the same general social principles found in other forms of truth seeking. Second, if scientific thinking heavily weighs certain kinds of knowledge acquisition based on objectivity (and thus is guided by certain kinds of logical strictures and values), nothing suggests that this form of reason necessarily trumps other hermeneutical, aesthetic, or intuitive forms of reason that play their own respective roles in the work of scientific inquiry.

If we understand that human reason exhibits diverse logics and that science is constructed from various interests, values, and modalities of knowing, then the kinds of analyses that might expose those contributions seem critical for a fuller understanding of scientific thinking. Once we pass through the door into a larger intellectual arena, we possess an expanded way of thinking about science's reason, which follows from the enlargement of a narrowly conceived objectivity and collapse of a rigid fact/value distinction. Further, once we understand a wider array of values as contributing to scientific knowledge and allow them their rightful place in the calculus of knowing, opportunity beckons for science to join a larger menu of concerns than its traditional twin roles of supporting technological innovation and mastery of nature.

On this view, we might replace the science/nonscience demarcation with another duality, putting science's pragmatic concerns on one side and science as part of a larger intellectual enterprise on the other. Science then resides on both sides of the divide. The pragmatism of scientific practice describes local realities containing various objects of inquiry, defining them as they are manipulated and used as tools towards finding new objects and the relations between them. A second dimension of scientific knowledge is, frankly, intellectual, expansive, and open to seeking its own place in a world framed by all kinds of knowledge and filled with human industry determined by several kinds of reason.

In this second domain, scientific inquiry readily admits its socio-political activity. Having become a major determinant of how we live and the social goals with which we grapple, scientific facts either subtly direct or ground a political orientation, or conversely, facts become articulated in a complex alignment within the particular social context in which they are employed. The public character of science as a contributor to understanding human nature, as a purveyor of rational deliberations about social policy, and as an exemplar of the values mediating public civility show how scientific findings and application depend on the value structure in which the facts are construed and applied.

Science as politics has taken form in "the boundary question." Sociologists of science have convincingly shown that the boundaries circumscribing science are at least porous, and some would argue, wide open. Multiple agendas are afoot in any public deliberation that might use scientific findings, and in scientific testimony, we witness science applied to human problems and opportunities, all embedded in a complex array of human values and intentions. Indeed, facts are the semantics of such deliberations, but the grammar conferred by the values that interpret and configure the words (facts) create the linguistic meaning. From this vantage, we clearly appreciate how the scientific enterprise supports liberal societies and why continued promotion of scientific research and education not only offers material wealth and economic prowess, but also serves as a bulwark of idealized rationality and the ethics that accompany it. Furthermore, the role of science in defining personal identity cannot be over emphasized. After all, our basic notions of human nature and the social character of society derive from the scientific corpus in both obvious and silent ways. That aspect of framing political identity suggests that we live in an ether of science in which every breath draws from the reality depicted by the scientific picture of the world and human mind. It behooves us to better understand that atmosphere.

To characterize science within such a large context, I have adopted a particular viewpoint: We live in a postpositivist age, characterized by skepticism about formal systems, so instead of formulating the *logic* of scientific discovery, we have settled for *pragmatic* descriptions of scientific theory and practice. Yet within this self-conscious appraisal, notions of objectivity and truth remain as guiding principles of scientific discourse. Holding to that balanced view remains the critical issue in contemporary discussions about the character of scientific practice and its conceptual grounding. I have four basic orientations: first, to avoid the excesses and dogmatism of either extreme of Science Wars commentary; next, to present a conception of reason that opens an avenue out of the interminable debates within science studies about the nature of objectivity and neutrality; additionally, to place science firmly within humanistic concerns that have too often been ignored; and finally, to offer a conceptual approach for understanding science as an evolving relationship between facts and the values that govern their discovery/manufacture and applications. Within this circumspect appraisal, we must defend science's rightful epistemological claims from the assaults of radical relativists in order to confirm that truth and objectivity function, *at least*, as working ideals. On this centrist view, truth and objectivity have lost their Platonic status and have been brought down to earth to reside within their sociological context, where they are employed as pragmatic tools.

In sum, I am committed to placing science within the humanistic context from which it originated. In that placement, my interpretation unlocks the interplay between science as an epistemology and science as part of a metaphysical construction of reality. By appreciating the wide reach of sociology of knowledge, we achieve a deeper comprehension of the scientific enterprise. *Science and the Quest for Meaning* thus presents a description of science from two vangage points: the instrumental and the humanistic. These are not necessarily opposing, or even in competition, but rather complementary. To disregard the original humanistic role of science distorts its character. By acknowledging the wonder of nature and a search for meaning as crucial sources of scientific imagination, we uncover a richer and more comprehensive picture of modern science. This older piece of the story, forgotten or too often ignored, finds its rightful place here.

Narrative Plan

My presentation requires no background in the vast literature of science studies, so the narrative is suitable for the general reader or undergraduate student seeking an introduction to this topic. A philosophical tack is taken here, so while the discussion is framed historically, much of what follows depends on explicating the underlying philosophies of science, which have dominated competing characterizations of laboratory research and its application since the nineteenth century. I have assumed that the reader possesses a basic knowledge of philosophical terminology, but certain complex issues (such as arguments about the character of realism, the role of teleology in biology, and various theories of truth) are explicated with discursive endnotes. Certain key figures—Willard van Orman Quine, Kuhn, Polanyi, Feyerabend, Hilary Putnam—serve as nodal points of the discussion, but a comprehensive description of positivism's history and its post-positivist successor lies beyond my purview. And to balance this philosophical survey, due attention has been paid to the contributions of sociologists in forming the dominant view of science presented below. Indeed, the "sociological turn" away from philosophical formalisms captures much of our contemporary understanding of science.

A broad survey, at the expense of a topical approach, has been adopted to track the developments in our understanding of how science is conducted and applied. This strategy differs from most commentaries, which analyze science according to particular points of view —logical, epistemological, sociological, and so forth. A more global overview arises from a deeper difference than diverging scholarly methodologies can reveal. In due course the sources of these divergent views and the significance of their differences will become evident.

Chapter 1 begins with a general description of the issues discussed in this essay. These matters include characterizing science (1) from a humanistic perspective, (2) as a postpositivist epistemology, and (3) as constitutive to the political life of Western societies. As mentioned, my interpretation rests on a reevaluation of the fact/value distinction that marked positivistic science, with an understanding of their complex interplay and changing relationships.

I follow this eclectic description with a more detailed discussion of each element as narrated through their historical development. Chapter 2 presents a portrait of the prevailing nineteenth-century philosophy of science—positivism—which most starkly set science apart from its older, broader philosophical concerns. The segregation of professional "natural philosophers" as "scientists" distinguished them from their philosophical colleagues, not only on the basis of their methodological differences, but also because of the experimentalists' explicit rejection of metaphysical concerns (or at least so they hoped). Their successes grew from the discarded subjectivity that had characterized so much of romantic science. From that position, metaphysics smacked of the subjective (understood as unsubstantiated). The positivists summarily threw the baby out with the bathwater. Sociological and intellectual divisions thus established eventually evolved into the Two Cultures of the mid-twentieth century. This history frames the criticism directed at the prevailing philosophy held by the defenders of science who continue to challenge the descriptions offered by contemporary postpositivist science studies.

Chapter 3 describes how the Two Cultures assumed a wildly different relationship with the radical critiques of science studies scholars during the 1960s and 1970s, when Kuhn, Feyerabend, and their followers stormed the citadel and shoved sociological analyses in the faces of their incredulous scientist colleagues. (The philosophical foundation of their appraisals had already been set by Quine about a decade earlier.) Differing visions held by postpositivist critics and defenders of science led to the so-called Science Wars in the 1990s. In chapter 4, we will review both the fruitful discussions and the headline-grabbing histrionics. In those polemics, two poles of opinion emerged: on one side, science studies critics claimed that science had no singular, idealized method, and that its discourse comprised many different, sometimes competing, forms of reason. Dominated by various kinds of constructivist depictions of truth claims and theory formation, the more radical interpretations presented science as little more than a rhetorical agonist field of political conflict. These "anti-science barbarians" (as they were called by some self-proclaimed defenders of science) represented a fringe of constructivist critics, but they brought upon the entire field of science studies the ire of laboratory scientists and their advocates. The radicals were accused of debasing the rationality that had served orthodox dogma of scientific reason for at least a century, and while much of the defense was justified, even the most modest constructivist claims were unjustly discarded.

As presented, these opposing positions appeared almost as cartoons, but those debates, whose hyperbole must be bracketed, were hardly trivial. This book explores their deeper significance. Indeed, beneath the theatrics, profound disturbances rose to the surface, and in the boil of that dispute, we discern a fundamental conflict in play over how to conceive of science and, even more, rational discourse. Perhaps still not fully appreciated is that the issues underlying these debates touch every sector of the public domain (e.g., education, politics, religion, and the law) where the reason that governs science has been applied.

With this background, chapter 5 presents science in its larger social and political contexts, where the interplay of facts and values is amply illustrated. The general notion of science as politics extends from government deliberations to personal understanding of scientific knowledge. With one eye cocked towards recent political events and the other towards a wider conceptual appreciation of the intellectual endeavor called science, we will explore how the fingers of the laboratory extend throughout its supporting culture and how its lessons are applied to every reach of the social. In short, the boundaries of science are indistinct and ever changing, which only reemphasizes how the applications of scientific knowledge often result in moral challenges of various sorts. In an extended case study, we will examine how the value structure of ecology, a scientific discipline, has been extended to environmentalism, a political and social movement. I will argue that biology incorporates nonepistemic values, which then allow the extrapolation from natural criteria to the moral tenets comprising environmental ethics. The two domains, epistemological and ethical, thus must be regarded as overlapping and informing each other.

Finally, the conclusion offers a synthetic perspective in which to characterize the multifarious activities we call science. I seek to rebalance the original pursuits of science, namely, the dual goals of mastering nature (i.e., development of technologies) and serving the pursuits of meaning and significance in human terms. That discussion restates science's broad humanistic commitments to redress the imbalance resulting from the first agenda subordinating the second. In order to clarify this issue we will consider the outline of a philosophical overview that places the pursuit for meaning again on a par with the project of finding truth. Extending the subject beyond "science and religion," we will examine how scientific knowledge becomes "personal knowledge" (for example, its impact on existential issues of self-identity or the relation of humans to nature). In this context, a more complete portrait of science as a theory of the real, a reality of our own, emerges. Drawing from science's own humanistic tradition, I am building an integrative program, one constructed from various forms of reason and the diverse faculties of knowing that make science, *science.*[4]

1

What Is Science?

The point is not to secure objectivity but to understand it.

Edmund Husserl, *The Crisis of the European Sciences*

How are we to understand science? I mean, *what* is science? Dictionaries offer succinct answers. For instance, my unabridged tome offers five definitions, each of which refers to knowledge: knowledge as opposed to ignorance, knowledge as a systematic account of nature, knowledge directed toward a specific object or phenomenon, knowledge obtained by a specified method or accepted scientific principles. Ironically, each are correct, but they miss a key aspect of what makes science, science. This book argues that while science is *knowledge*, a very special kind of technical knowledge, it is more. It is also about *meaning*—how scientific findings and theories become personally significant in the terms in which we think of ourselves as part of the social or natural universes. While obviously important, that aspect of the scientific endeavor receives little attention. In academic discussions, such considerations have fallen out of fashion. The dictionaries omit the missing element, focusing on the character of knowledge in a narrow sense; and this key humanistic element typically is either forgotten or never appreciated by scientists qua scientists, who focus on their technical pursuits. Indeed, in my experience only the rare investigator ever entertains a conscious thought about the "more-ness" of her research—the larger reasons she pursues her tasks. But a deeper reason underlies the segregation of science from wider metaphysical considerations.

Science, at least since the early nineteenth century, developed in competition with another idea of what science accomplished, or at least tried to achieve. Let us call the two visions *Ancient* and *Modern*. The older version still resides with us, but in disguised forms. At times during the past two hundred years, it has reared its head, looked about, and then sullenly retreated to the dusty bookshelves, where historians and philosophers sometimes wander. Its shyness only reflects the sorry state of its lost standing. The Ancient notion has been in retreat since the Renaissance, and some would argue that "Ancient Science" distorts not only what in the present we understand science to be and to do, but also changes the playing field of what the Babylonians, Greeks, Romans, Chinese, and Arabs were doing in their own investigations of nature. On this view, science as we know it emerged only in the modern era, and more, "modern" may well be synonymous with "science."

For the ancients—and Aristotle is the key figure in that history—science was a means to classify the natural world in terms of "natural kinds." Everything possessed an essence, that which made it an individual. This seems a reasonable place to start. After all, as we watch children discovering their world, they ask, "Is this a brown?" The parent corrects by providing another class of thing: "No, this is a dog," and the child must adjust her notion of brown's essence. Classification places humans within their environment, for it is the first step in navigating the natural world. But classification alone hardly suffices for science. It serves as the beginning, but only the origin for other, more ambitious pursuits.

We must assuredly credit the ancients for discovering certain scientific principles, those we would now call simple mechanics. However, the various theories explaining those phenomena, from our modern point of view, failed miserably. Things possessed essences and the nature of things—birds, meteors, tides—was to follow their true character. Physical forces were imbued with some version of vitalism; the world was integrated by a life force, and that life force explained *everything*. What humans experienced subjectively then was projected onto nature, and that was explanation enough. Given certain premises pulled out from human experience, the ancients created a world order. Experimentation did not exist, certainly not as a test of a predictive model or theory. And facts were determined by individual observation, which for the most part required no test of veracity by a committee of peers. Authority ruled, not in the tribunal of empirical knowledge, but rather in the abbeys of religious opinion and dogma.

Ecclesiastical authority reigned in the premodern era. The Church determined theology, a philosophical doctrine about the divine, and that body of knowledge was derived from both revelation and scholastic argument. Here we find the first clue about our initial question. To the extent science was science during the Reign of Religion, it served a metaphysical purpose: examining God's grandeur revealed in nature. What is the nature of the universe in which humans mysteriously find themselves? How might life after death be understood, given what we know about earthly living? Where is humankind placed in the hierarchy of nature? And based on this last query, again extrapolating from human experience, those who practiced early science were to finally ask what God's laws are—laws that not only applied in the moral domain, but also those governing the natural order.

For science and religion to align themselves in balance ultimately depends on recognizing the legitimate rationality of each and the creative exercise of finding mutual accommodations. That challenge has not always been achieved despite seemingly endless attempts to maintain stable relations (Marcum 2003, 2005a). We will consider a case example in detail below, but suffice it to note here that the various celebrated breakdowns of mutual tolerance testify to the fragility of equilibrium. Indeed, today, while many still seek to accommodate the claims of religious and scientific worldviews, it appears that science dominates such discussions and religion has been put in the position of finding its own place in relation to the reality science has presented. Following the historical development of science's material and theoretical successes, the scientific enterprise has achieved independence.

Once the question of universal law emerged, science began its course towards its modern identity, one based on a new kind of empirical argument (Olson 1991). If our contemporary society accepts an authority, it is the authority of scientific truth. That authority rests on the stature of empirical facts, and facts have very particular standing in the scientific lexicon. Much of what I will be discussing concerns the character of facts, and how the concept of "factual" is hardly simple (Poovey 1998), nor for that matter, is the orthodoxy of "scientific method" (Gower 1997), the very history of which goes hand-in-hand with the vicissitudes of defining facts. Whether in the laboratory, where facts are discovered and made, or in the social world, where facts are used in policy debates, we will see that their standing is persistently contested. Whether a fact is a fact always remains an open question, and how to interpret them, either as supporting or refuting a theory, or claiming valid judgments about the world, remains in the arena of disputation.

This circumspect view of facts has arisen only recently. During the nineteenth century, facts (determined by proscribed objective methods) claimed their standing by denying subjective bias in observation or report of scientific inquiry. In reaction against the authority of personal confessional, the human observer was to recede into a "subject-less" recorder (Fox Keller 1994). To the extent the investigator gleaned facts from her observations, she did so as a machine among other machines. Her instruments were, to be sure, extensions of her own perceptions, but she recorded those observations solely on the basis of some mechanical measurement, a direct transmittal of nature's measure to a fact-recording device. That machine, a coupled observer-instrument, became the paragon of scientific virtue: no subjective bias could be introduced. And facts, the products of that process, became sacrosanct. This conception of scientific study, its rise and fall, fills my story, but before disabusing belief in what has been called the Legend (Kitcher 1993), let us briefly review the hard-won struggle over different kinds of knowledge that has played out historically in the larger social context.

Science always has had a political agenda, political in the sense of claiming its legitimacy in the pursuit of its own influence and receipt of social and economic resources. When Francis Bacon promoted empirical research and obtained the Crown's financial backing in the sixteenth century, he did so on the basis of the promise of material gain—economic and military most directly. However, a deeper and more profound political issue was at stake: epistemological authority. Knowledge based on empiricism and its effective application for material gain effectively challenged the power of the Church. While science abdicated any formal commitments to define religious beliefs or the existential status of human beings, nevertheless, with the re-alignments generated by new conceptions of the natural world and the standing of humans in that universe, older ecclesiastical teachings were directly confronted and the Church weakened. The equipoise of radically different epistemologies could not be maintained and the political consequences, in the widest of all possible meanings, cannot be exaggerated.

Since the Renaissance, scientific interpretation has challenged religious authority. And in that struggle, science has found itself, willy nilly, aligned with humanists and then secularists, who were able to challenge religious dogma with scientific evidence (e.g., the age of the earth, evolution of species, etc.). Secularists considered that political struggle, until the Bush administration, essentially completed. The fundamentalists apparently had been vanquished. After all, by the mid-nineteenth century, God's funeral was well underway and by fin de siècle, secularism had claimed its laurels. But the battles hardly ended then, and our own era has witnessed the struggle in different guises. Let us explore a recent case study, which I trust will illustrate the major issues still in play.

Reason in Dispute

During the week before Christmas 2005, Judge John E. Jones III, sitting in the Federal Middle District of Pennsylvania, ruled against teaching a new form of creationism in the public high school. The case arose from a suit brought by parents against the Dover school board, which had instructed teachers in 2004 to read a short statement about the inconclusive status of neo-Darwinian evolution theory and suggest that intelligent design might be entertained as an alternative explanation. After a long trial that delved into the nature of scientific theory and the questions of what constituted scientific knowledge, the judge ruled intelligent design was a ploy to bring religion into the classroom and accused certain board members of duplicity. Judge Jones only confirmed what the voters had already accomplished by pushing the errant board members back to church.

The courtroom drama riveted the country, some comparing it to the Scopes circus of 1925, when Clarence Darrow confronted William Jennings Bryan in the famous Tennessee "monkey trial." The 1960 movie *Inherit the Wind*, so well enacted by Spencer Tracy and Fredric March, captured my own imagination as a youngster, and then, as now, I was fascinated with the arguments about God's presence or absence in nature. I can well understand how religionists regard nature with awe, and to find coherence and, perhaps more importantly, meaning in the cosmos, they cannot abide placing their god outside his handiwork. If God is present in their daily lives, why should he be omitted from designing the greatest of creations, human intelligence? After all, the Bible describes how Adam was made in the image of God. Accordingly, God's intelligence, like our own, must have some engineering capability dwarfing even our wildest conceptions. True believers maintain that orthodox scientists are blind to a deeper Reason because they have yet to see the Creator's fingers at work. So what looked to modern-day Darwinians as only a contingent, blind evolutionary process, is in fact only understandable as an act of deliberate design.

The Dover case took on a special luster during the summer of 2005, when Cardinal Schönborn wrote a controversial op-ed piece in the *New York Times*. He claimed that he was protecting "rationality" against an ideological science:

> The Catholic Church, while leaving to science many details about the history of life on earth, proclaims that by the light of reason the human intellect can readily and clearly discern purpose and design in the natural world, including the world of living things.
> Evolution in the sense of common ancestry might be true, but evolution in neo-Darwinian sense—an unguided, unplanned process of random variation and natural selection—is not. Any system of thought that denies or seeks to explain away the overwhelming evidence for design in biology is ideology, not science. . . . Now at the beginning of the 21st century, faced with scientific claims like neo-Darwinism and the multiverse hypothesis in cosmology invented to avoid the overwhelming evidence for purpose and design found in modern science, the Catholic Church will again defend human reason by proclaiming that the imminent design evident in nature is real. Scientific theories that try to explain away the appearance of design as the result of "chance and necessity" are not scientific at all, but as John Paul put it, an abdication of human intelligence. (July 7, 2005)

The slippage is evident: Schönborn propels his metaphysical reason, that which supports God's cosmological purpose, into the epistemological domain, where the preponderant scientific interpretation sees no design (and, incidentally, makes no comment about God's presence or absence). In other words, he conflates theological reason with scientific reason and trespasses the boundaries as if there were no difference. (Schönborn's position is, of course, disputed within the Church, and other options obviously exist, but his orthodoxy frames my discussion.)

Schönborn chose to ignore the lesson Immanuel Kant taught more than 200 years ago—how reason must make way for faith. In the *Critique of Pure Reason* (1787/1998), Kant advised how to circumscribe objective knowledge and leave belief to reside beyond science's horizon. His formulation provided a model by which science and religion might coexist secure in their respective domains, by alerting the natural philosopher (now called a scientist) not to probe into areas to which scientific method had no ready access. He profoundly understood that science would not ask for, and thus would not offer, a basis for religious belief, one way or another. In a sense, the question of God became moot. Science erected a neutral picture that tilts one way with God, and another without him, but which way the cosmos leans is dependent on individual choice and belief.

Following Kant, science may allow a divine presence, but only one consistent with the best scientific interpretations. After all, being a scientist hardly precludes religious commitments. The issue concerns the standing of belief, as various forms of *knowledge* must be differentiated from religious *convictions*. But Kant's pluralistic option threatened those who could not claim the same kinds of certainty science exhibited, one that employed a different kind of rationality and a different basis for judgment. Of course, science's worldview is not necessarily incompatible with a divine presence, but protecting free inquiry and open interpretation remains a challenge. The current conflict over evolution in particular, may be charted as a set of concentric circles: disputed claims of biology; differing views of science and its role in our society; arguments over the limits of the state; and, most generally, the status of secularism and the Enlightenment project.

And here we come face to face with the secular-religious tension in its starkest terms: Schönborn's metaphysics demands divine intervention, and he would employ reason to support his belief. Rather than provide divine presence and teleology with its own rationality, he insists on projecting his faith into the natural world. In short, because his reasoned theology apparently can not accommodate neo-Darwinian blind evolution, Schönborn must reject dominant scientific opinion, namely the evolutionary findings arising from a nonteleological, materialistic theory that specifically rests on a denial of design, and, consequently, a displacement of a master divinity. Indeed, contemporary evolutionary theory could be interpreted as rejecting major assertions of Christian theology, and much else, for each form of modern biology—from molecular biology to sociobiology, from the heart's beating to the brain's functions—rests on the utility of chance events. For those who insist on God's immediate presence, reason dictates otherwise. Reason then becomes the tool that some theologians use as a kind of universal solvent for dissolving problems without acknowledging that it is not *reason* that is in dispute, but rather the metaphysics in which reason functions.

The question of whether intelligent design might take its place in the scientific menu does not strike me as particularly interesting at this point.[1] We have witnessed endless and convincing rebuttal, but what intrigues me, and the set of questions upon which I will focus, concerns the character of reason and the characterization of scientific reason in particular. The key to the Cardinal's position concerns a definition of rationality, since he claims a higher rationality than the neo-Darwinists. But reason is not at issue inasmuch as the logic by which he argues is perfectly consistent within his own system of thought. Instead, the trajectory of the quarrel takes the disputants into diverging paths because of the presuppositions each holds. Each declares his basic axioms and then follows the logic of reason to some conclusion. Schönborn begins with God's presence and purpose (humans created in a divine image and living for divine resurrection) and then sees God's design imposing an omniscient hand on evolution (to achieve holy ends) as consistent with another basic premise, God's limitless will. The counter position begins with blind evolution, sees no design, and then remains agnostic about the Divine's role in the evolution of life forms. Even if both sides acknowledge the difference in their respective assumptions, the debate goes unresolved because reconciliation at this level is not based on rational argument, but instead on the *beliefs* underlying each party's initial assumptions.[2]

Both sides of the debate claim a rational discourse, and indeed, intelligent people espouse intelligent design, but given the presuppositions of each system, the conclusions of the respective positions are irreconcilable. The incompatibility stultifies argument because presuppositions are, as R.G. Collingwood described them, the assumptions and guiding precepts that are closed to further analysis or revision (Collingwood 1940). They are the bedrock of the conceptual apparatus they support. Start with different presuppositions, and logical progression will bring the disputants to very different ends, as the intelligent design case exemplifies. So the public drama is not about science per se, but about the metaphysics in which science functions. The classic examples are the religious disputes arising from Galileo's astronomical findings and Darwin's theory of common descent. In each instance, a religious orthodoxy disputed the science. Galileo's case has been settled, but Darwin's still lingers, not in the particulars of evolutionary findings but in the meaning of those findings, as the Cardinal has shown. Indeed, scientific facts are not at issue, but rather their interpretation, so that we should recognize the instrumentality of reason: Science may be used by anyone; its technology applied for diverse social pursuits; its knowledge perhaps designed for one purpose, applied to another; its findings interpreted to support one metaphysics, or another.[3] When the fossil record is placed within a fundamentalist reading of the Bible, a "metatheory" has supplanted the scientific one. We will not settle the matter by argument, rational or otherwise.

However, a deeper issue lurks beneath the merits of the Cardinal's pronouncement. I argue that science has been unfairly indicted with the responsibility of ousting humans from a sheltered niche where we resided unique in nature as privileged creatures in communication with God. According to the critics, metaphysical disjointedness is laid at the feet of an imperialistic science that not only defines nature and human beings in an anti-spiritual language, but also calls into question other modes of knowing the world, ourselves, and the Beyond. But the accusations represent a profound misunderstanding. Science does not, cannot consider religious claims. Since science makes no attempt to address or listen to God, the question of whether the divine exists or not is simply off the scientific agenda. God resides beyond scientific discourse, which simply means that existence is mysterious enough to make room for both knowledge *and* belief. In a sense, God is besides the matter. (This "complementary" view of the nature of the science/religion relationship has been exhaustively described, e.g., Brooke 1991; Barbour 1997; Ferngren 2002.)

While scientific theory is neutral regarding the divine, understanding the mystery that lies at the heart of the scientific query originates with the very same religious questions that evolved into philosophical ones. What is the world? How is it organized? Where do humans fit into that universe? What is distinctly human? Science presents cogent "answers" in its distinctive voice. Indeed, science cannot escape its "intention," that is, the abiding *human* questions that direct its inquiry. The "view from nowhere" (Nagel 1986) not only remains an impossible aspiration when science is conceived in these human terms; the presumption also radically misconceives science's own commitments. While the terms of engagement had been radically altered by the nineteenth century, the original metaphysical inquiry remains embedded in the scientific enterprise as a second order activity. *First order* refers to the direct industry of science, its epistemological project defined narrowly. *Second order* refers to the interpretations and applications of scientific findings and theories. Scientists in their technical work deal with first order business; all of us ponder second order issues. Second order answers have been asserted by religionists and secularists for their own respective purposes, and in their ensuing debates, they have used scientific theories to support radically different metaphysical positions. The science (as facts) is not the issue. The *interpretation* is. In this sense, science fulfills an instrumental function far beyond its material applications.

Thus science, from an enlightened religious point of view, is an instrument to perceive the divine. The ingenuity of scientific investigations exposes the wondrous workings of God's hand. The more we understand these investigations, the more we might appreciate the sublime coherence and intricacy of the natural world and its unfathomable reaches of space and time. That articulation, for the religiously inclined, reveals God. The religious integration is thus closely linked to the aesthetic, one that draws from notions of the sublime.[4]

So instead of the negative project of rejecting fundamentalist arguments, science may be employed in the positive endeavor of translating its own picture into terms that appeal to subjective needs. On this view, science not only provides the basis for technological advances, but answers to its deepest commitments of exploring nature as a response to our metaphysical wonder. Humanism leaves the chore of defining significance and meaning within a human construct. That challenge lies at the base of the conflict between secularism and religious ideology. In a sense, Nietzsche's challenge ("God is dead!") remains an abiding unresolved question: Can, or even should, humankind define its cosmos? Beyond naturalistic explanation, can the values that govern society be truly based upon, or even derived from, human deliberation? Can we successfully assert our own significance? Can we meaningfully exist without divine revelation and live in a world navigated and created by human intentions and will? These questions have rested at the heart of the secular enterprise throughout modernity. Indeed, they largely define the humanistic project, and when liberal society is confronted by such expressions of discontent as in the Dover case, we are reminded that, for a vast proportion of Americans, the world science presents cannot provide meaning that satisfies their existential and religious needs.

Scientific findings by themselves offer no meaning in a humane sense. Interpretation orders those facts into a construction that, in the final analysis, is an attempt to place objective knowledge about the natural world within the broader dimensions of human experience and subjective needs.

That process requires some "framing"— aesthetic, spiritual, and moral. The fundamentalist legitimately aspires to integrate a scientific picture—evolution—with deeply held religious commitments. But this is hardly a problem unique to them. Many secularists (and free-thinking religionists) also seek seamless connections between a materialistic universe governed by laws that have no personal enchantment with the various dimensions of subjectivity. They do so without invoking divine intervention in the particular evolution of humans. All of these groups share the same problem but reach different solutions. Indeed, the aspiration to understand individual identity and "place" persons within the various natural and social worlds they inhabit (psychologically, sociologically, spiritually, and so on) seems a universal characteristic of human life. Viewed in the most general way, human reason apparently has a basic property (one demonstrated by myriad psychological and cognitive studies) of seeking integration of experience, unity of belief, and coherence of understanding (Thagard 2000). Kant described this integrative function as reason seeking its "unification" (Nieman 1994). Freud discovered numerous defense mechanisms to hold the psyche together. Cognitive scientists have demonstrated the ability to screen out or forget data or experience conflicting with more dominant belief (Hookway 2002; Fauconnier and Turner 2002). And metaphysicians jealously guard their presuppositions to hold their world together.

We seek to understand how various reasons effectively knit the world together, with science serving as a paragon of a certain kind of knowledge, but only one of several in the employ of this integrative function. To assert the legitimacy of different rationalities is not to advocate relativism, but to acknowledge that there is no single epistemology that may lay exclusive claim to all domains of experience. Each form of knowledge explores and then defines the world according to its own means. The common mistake admits no limits of a particular mode of reason. Rather than seek a metatheory (or version of reason) to encompass each mode of knowledge, respect for intellectual and social experience of each practice must sustain the rightful claims of all by appreciating that once we reach metaphysical strata there is no *relative* merit in scientific versus other kinds of reasons, for example, that governing religion. So the tack taken here is guided by a sighting of reason; the winds are coming from starboard; we require a steady compass to hold our course. Let us proceed accordingly.

A Metaphysical Quandary

Knowledge offered by scientific investigations resides at several levels, and although the material benefits testify to the success of the scientific worldview, characterizing those technical applications are only one aspect of understanding science's product. So beyond the material gains, we might well ask the more general question: if the reality depicted by scientific knowledge may only be grasped indirectly, in other words, through technical means expressed in esoteric language with strange modes of logic and an obscure history, then what can a nonspecialist *know*? Or for that matter, how does the expert place her technical slice of reality into a comprehensive worldview? The scientist faces the same challenge of processing and integrating her own knowledge, and quickly becomes a layperson when she wanders off her beaten path of study. According to this egalitarian view, each of us watches the reality depicted by scientific discourse much as we might enjoy a movie: we view a version of the real, some segment of the natural world pictured with varying degrees of detail. We do not necessarily experience that reality directly, but rather understand it through some intermediary or derivative interpretation. Making knowledge personal—meaningful or significant—remains then an individual's predicament. The implications of standing off stage, as it were, have deep repercussions as humans become voyeurs of a world in which they live. In short, how to reenchant the contemporary worldview—namely how to derive meaning and to find significance in a world devoid of human value—presents a perplexing query.

Romantic fears and disclaimers that such an integrative mission might be doomed to failure profoundly influences Western consciousness. Mary Shelley's *Frankenstein* lurks everywhere, a looming hulk threatening us all. Faust suggests a pact with the diabolical, and whether the devil lives within our midst or within our own hearts, the danger is the same. Suspicion of the scientific mission may be traced from the eighteenth-century infatuation with the Noble Savage (expressing lost innocence) to Nietzsche's celebration of Dionysus (declaring an aesthetic liberation from a restricted rationality).[5] Each attack on the insidious character of scientific industry expresses a deep-seated and powerful sentiment of technological progress overpowering an essential humane component of the Western psyche. On this view, despite the obvious fruits of scientific labor, a damning indictment stands for dehumanizing industrial cultures (Marx 1979).

The pitting of modern science (coupled to a voracious technology) against the pastoral (the innocent and good) has created a broad cultural conflict, emanating from a deep chasm in intellectual values and *Weltanschauung* dating at least from Jacques Rousseau (Marx 1964). Consider this indictment by Edmund Mishan:

> Like some ponderous multi-purpose robot that is powered by its own insatiable curiosity, science lurks onward irresistibly, its myriad feelers peeling away the flesh of nature, probing ever deeper beneath the surface of things, forcing entry into every sanctuary, moving a transmuted humanity forward to the day when every throb in the universe has been charted, every manifestation of life dissected to the nth particle, and nothing more remains to be discovered—except, perhaps, the road back. (1967, 144)

This epistle cannot be dismissed as idiosyncratic. The power and urgency of the antiscience lobby punctuated the twentieth century, and the current chorus has many expressions, ranging from the fears regarding nuclear power, to the potential untoward effects of genetic engineering, from the disputes about human evolution, to the status of the fetus. In the nondifferentiated dismay with the excesses of technology, science and its uses are generally not separated. But even when science and its applications are distinguished, a persistent complaint revolves around the "de-naturalization" of nature. Accordingly, unified nature has been torn asunder by a reductionist and radically objectified science that cannot put the fragmented parts back into a coherent whole. More, science's reduction of nature to an object of study, has been indicted with radically altering humankind's intimate relationship with the natural world.

Hostile commentators too often have failed to differentiate the purported crimes of science from the social uses of technology (Proctor 1991). The efforts to discern the complex political and economic forces guiding science, along with the more charged task of adjudicating the indictment of scientific callousness, have resulted in ongoing controversies that are unavoidably ideological in character. So while we might simply disregard Luddite critiques, a nagging insight lingers. Much of modernity's history is science, and the unresolved status of its objectified, universalized worldview vis-à-vis the subjectivity of the singular ego remains a critical matter.

The subject-object dichotomy lingers well beyond the artifices of research and extends to the very core of our standing in nature. In short, some integrated metaphysics governing humankind's place in nature has been displaced for many by a metaphysics of alienation. Needless to say, adopting this alienative perspective, leads to a nihilistic understanding of human's existential status. One need not subscribe to this line of thought to still acknowledge that the scientific worldview has presented modernity with a major challenge of defining a metaphysics that successfully integrates objective pictures of nature with the personal world in which humans live as social and psychological beings.

Martin Heidegger's provocative comment, "science is the theory of the real" (1954b/1977, 157), points us in the direction we will follow to tease apart these matters. He might have offered a more expansive definition, such as "science is *a* theory of reality" or "science is the *quest* for reality" or something allowing for other worldviews. He did not for a very specific reason: he wanted to jolt his reader to acknowledge that science has become the dominant way of understanding reality, and, more to the point, other ways of knowing have lost their standing. Heidegger purposefully assigned the scientific worldview a firm hold on what constitutes reality. He was a cagey fellow, and he no doubt took some delight in his attempt to entrap the unwary with his irony. More, he wanted to scandalize those who vouched for science's authenticity by declaring that picture both incomplete and distorting. That is a complex (and notorious) story, but the laconic dismissal (yes, science gives us the "real," but that picture is not really *real* or even truly interesting) points to the dilemma that frames my own consideration.

So Heidegger set the terms of engagement: if science is the search for the real, what does *real* mean and to what degree that scientific reality presents a comprehensive worldview? Or put another way, if science is the theory of the real, then what is it that remains for other ways of knowing or world-making? What is included, and, more importantly, what is left out? And considering that which is omitted, what is the cost of its loss? A broadened position regards science as ruling or defining only one domain of human knowledge and other ways of knowing are appropriately applied to discerning different aspects of personal or subjective experience. In other words, objectivity does not solely command our picture of reality, and, indeed, it can only contribute certain components to the complex mosaic of our experienced world. If one assumes this critical attitude, then the way of conceiving what science does and endeavors to do may radically differ from accounts of those who remain satisfied with the picture of *the real* science bestows. In short, Heidegger skeptically asks, what is the status of subjective experience in a world dominated by scientific realities?

Prior to the nineteenth century, such a discussion about the nature of science as an intellectual endeavor among several competing modes of knowing did not exist, at least not as a clear alternative to positivism. The rise of positivism not only redefined the character of reason and knowledge, it simultaneously redefined reality in its own terms, Approaches which might be posed in aesthetic or spiritual terms, not only were irreconcilable with a positivist philosophy, but would have to accommodate themselves to this new philosophy of science. The worlds for which science did not account were not conquered so much as bypassed. While science's epistemological hegemony affected ethics, religion, and of course, the human sciences, which in turn deeply influenced social policies of every kind, the most profound influence was on metaphysics. Note, when Heidegger observed, "science is a theory of the real," he qualified the epistemological standing of science by expanding the definition of science as also encompassing *metaphysical* concerns. Indeed, the latter are primary. Reality, namely, a philosophical understanding of reality, comprises the métier of metaphysics. When Heidegger makes what appears to me to be a transparently obvious observation, he immediately shifts the playing field for philosophical discussions.

Heidegger distinguished what science does (and the products of its doing) from the primary philosophical project to which it is committed—presenting a theory of reality. Indeed, scientific theories are the foundation for ordering phenomena and thus serve as the philosophical scaffolding for the entire enterprise. (Theory is being construed loosely here.) Accordingly, scientific investigations, the methods and factual products, are not the sole ends of the venture.

Beyond an epistemological enterprise, a more fundamental pursuit orients research, namely, the development of theories and discovery of laws that account for a distinctive way of portraying reality. In short, science produces epistemological as well as metaphysical statements about nature. With that understanding, the primordial origins of scientific inquiry, inspired by the wonder of nature, are appreciated as a "theory of the real" that makes this ancient metaphysical calling basic and *constitutive* to contemporary science.

From this metaphysical vantage, science encompasses an enlarged agenda, one in which the scientist not only seeks to master nature, but also to address the original queries of the earliest philosophers: what reality is, who we are, where we are, and maybe, just maybe, how we are. Moving past the technological achievements that we derive from scientific progress, these deeply human concerns legitimate science in a context beyond material needs, one that we might dub spiritual, aesthetic, emotional, existential, or just plain humane. Each of these questions arises from the conundrum of self-consciousness, a consciousness of ourselves in an alien and strange world.[6] With self-awareness, people of all time have asked these same basic questions, and science, for better and for worse, has provided its own unique answers.

We may lump these matters together as entirely outside of science's concerns. In some sense, that is a correct assessment, but only if we characterize science narrowly, as nothing more than an epistemological program. However, when we consider science as a broader form of philosophical inquiry with deep metaphysical commitments, then, the positivist program collapses as facts move beyond the laboratory to help construct worldviews that go well beyond science's distinctive epistemology. Indeed, facts are always interpreted and extended within larger contexts, and "interpretation" easily slides to "meaning." For instance, what is genetic determinism and how does it affect moral responsibility? Or, given the expected rise in carbon dioxide levels in the atmosphere, how would changed economic policies impact on social welfare (which, of course, depends on defining communal good)? Or, accepting current theories of evolution, where does a divinity reside, and what might it do? While science cannot directly address such questions, it remains an active participant in those deliberations. We could say that science frames the question and sits at the table as the moderator of the debate, albeit not always as a neutral partner. After all, scientific results by themselves cannot provide meaning, but the very character of the scientific picture often determines, and always informs, the derived answers. Whoever first quipped that "epistemology drives metaphysics" was profoundly correct. Science, as an epistemology, remains inextricably coupled to the existential questions that define human self-knowledge and understanding. That coupling deserves special scrutiny.

As we explore the *idea* of science, we must consider two competing conceptions: science as a tool to promote human well-being, namely an intellectual and technological enterprise to understand and control nature, and science as the framework for building existential and metaphysical formulations. In the first case, objective knowledge is applied to make things or propose generalizations (laws, hypotheses) about nature. In the second case, facts, laws, and scientific inferences are translated into personal knowledge to place humans *in* nature. Earlier I referred to these as first order and second order concerns, respectively, but I did not mean to imply that "first" takes precedence over "second." Each has its mandate, and while the domain of the first has a rich descriptive literature, charting the horizons of the second requires rigorous reiteration. I maintain that, not only do humanistic interests influence science, they lie embedded in the very foundations of the scientific venture. The quest for reality is, in the end, *our* quest for understanding, whereby the individual knower must ultimately process the universal. Obviously, material advancement is a crucial aspect of science, but the humane program, its original and abiding call, defines the commitment of discovering the world *for us*.

Technology represents the most obvious use of science to develop human industry. "Industry" does not refer here so much to material culture as to the more general understanding of industry as the systematic labor to create *value*. While science generates vast material wealth, its technology and resulting mastery of nature fulfill only a part of science's industrial agenda. On this view, science becomes instrumental in several senses, as an authoritative instrument for *describing* nature, as a powerful instrument for the technical *mastery* of nature, and as a personal instrument for *understanding* the world and navigating it. By focusing exclusively on description and mastery as critical tools of the modern mind, we neglect the complementary contributions the scientific enterprise makes towards a metaphysical orientation.

Part of the problem of integrating science's epistemological orientation and its larger metaphysical influences originate with positivism's ways of knowing. For positivist science to effectively achieve its goals, the investigative agent must stand back from nature and observe, ostensibly from a view from nowhere. The subject vanishes, and that vacancy depends on the Cartesian division between *res cogitans* and *res extensa*. Humans no longer reside in nature, but rather step out and look *at* the universe. Descartes thus defined modern consciousness arising from splitting mind from body, and the positivists simply built upon that platform. Their science exemplifies humans peering at the world neutrally. With this stance, a rift in experience divides subjective inner experience from dispassionate objective observation of the world and others. Held as a truism, science's reason appears radically different from personal judgment, which operates in the ethical, aesthetic, and spiritual domains to mediate subjective experience.

Given the Cartesian construction, this putative absolute division of subject and object provides science with its epistemological strength and its concomitant metaphysical dilemma. The strength is the objectivity conferred by the ability to detach observation (the observer) from the object scrutinized. The dilemma derives from the existential quandary that arises from separating the human subject *from* the world to become a dispassionate witness *of* it. This issue leads to the very roots of modernism, where the observer remains outside the picture she watches, yet is compelled to refract nature in human terms. And from this understanding a series of questions presents itself: Are we fated to peer forever at the world self-consciously, knowing that we are spectators of nature as well as of ourselves? Can we bridge the division of self and the world? How does the objective stare become personal? Since science's worldview, its metaphysical agenda, is unfinished in such terms, how do we complete it in ways that signify and make meaningful the picture presented? Specifically, if science reveals a world without value, and value is entirely human derived, then how are we to interpret nature in meaningful, humane terms?

Integrating the scientific worldview into personal experience and daily life beyond the direct material influences of technology and social policy challenges us to ask how science joins an assembly of different forms of knowledge and different ways of knowing (Toulmin 2001). Beyond how we might understand science as an intellectual enterprise or as a cultural institution, we must reflect on the implications of having a worldview framed by science. *Worldview* refers not only to the picture of reality science bestows, but also to the manner in which scientific thinking profoundly affects *the way* one perceives the world and oneself, individually. To approach this general issue, we must consider how a translation occurs between the objective picture of the world and the meanings by which we signify that world. I am referring to an understanding of science's own rationality in relation to other kinds, and in that comparison, describing where we might place the personal, subjective ways of knowing. Indeed, how might we deliberately conjoin human-derived, human-chosen, human-centered values with those objective values that we so commonly understand as irreparably separated from these origins?

The placement of scientific knowledge within a "web of beliefs" (Quine and Ullian 1978)—public and personal—requires faculties that draw upon creative resources to integrate the objective world science presents with other social and subjective values. The ongoing creativity required to make the world whole, to cohere, to hold meaning and significance, attains a certain hue when science is idolized as a false divinity, some *thing* separated from human industry, as opposed to functioning as an instrument of human imagination. The challenge emerged in the early modern period, became more clearly articulated in the romantic era, and is now renewed by postmodernism's suspicions of enduring structures, particular logics, espoused perspectives, cherished ideals, and universal values. Indeed, postmodernism embraces a neoromanticism in its collective endeavor to recapture a world oriented by the multitude of human interests and human values, a world characterized by pluralism, legitimization of diverse forms of reason, and an appreciation of the communal character of individuality.

The West of today is no less challenged than earlier romantic critics by a hegemonic reason divorced from the realities of human need and value. We too must consider how to fit the scientific mode of knowing within the broader humanistic agenda. Here, I argue that we cannot regard science solely as some kind of separate activity for studying the natural world. Rather, the scientific worldview has assumed its dominant place in contemporary society through its constitutive status of defining *human* realities. From this point of view, we must understand the technical mastery of nature not only as a Baconian fulfillment of material advancement and mastery of nature, but also as a response to the primordial, deep stimulus of metaphysical desire to know the world and place ourselves within it:

> The ultimate basis on which all our knowledge and science rests is the inexplicable. Therefore every explanation leads back to this by means of more or less intermediate stages, just as in the sea the plummet finds the bottom sometimes at a greater and sometimes at a lesser depth, yet everywhere it must ultimately reach this. This inexplicable something devolves on metaphysics. (Schopenhauer 1851/1974, vol. 2, 3)

If understood in this fashion, as already discussed, the apparent conflict between science and religion cannot be about metaphysics per se because science itself originates from the same questions that inspire both worldviews. The criteria of truth, the methods employed, the political authority of each, radically differ, but the constellation of primordial human wonder holds both approaches to the same line of inquiry.

So, let us consider a conceptual schema, which ties together this discussion and thereby also draw some conclusions: Good reasons abound as to why the original, deeper meaning of science has been deliberately obscured. One constellation of answers revolves around the practical results of science's epistemological enterprises. Those successes seem to depend on the eclipse of subjective and metaphysical contaminants that would conspire against the ideals of objectivity and neutrality. Much merit supports this thesis. The practitioners and philosophers of science embraced positivism and thereby discarded the assembly of metaphysical and existential questions that their science either totally ignored or chose to answer in its own, highly circumscribed fashion. Simply put, some questions were not answerable and therefore irrelevant to the scientific agenda, and others were recast to conform to its worldview. By the 1920s, positivist philosophers were even disallowing such metaphysical questions as "nonsense" and not worthy of analytical deliberation at all (Frank 1949; Ayer 1952).

Of course, metaphysics cannot so easily be dismissed, and even the positivists embraced a metaphysics, which would support their own philosophy. However, to be fair, they were addressing the metaphysics of religion and other belief systems whose forms of knowledge rested on opinion or revelation, as opposed to empirical facts. Some would maintain that the positivists asserted a new orthodoxy and flung the pendulum too far from center, but we should not lose sight that they regarded themselves as fighting the same basic battle marking the legitimacy of scientific thinking against the religious: Simply, they pitted science's empirical objectivism against a religiosity that rested on metaphysical speculation. Nothing less than the truth was redefined in the process. Referring back to our earlier discussion, the ongoing debates about the relation of science and religion are directly traced to this fundamental standoff over what one can say about metaphysics if one cannot prove a metaphysical statement scientifically. The answer, failing scientific scrutiny, falls into the domain of personal belief, and such belief becomes an individual truth category—subjective, unproven, and therefore valid only on its own grounds. Claims to some universality must be abandoned, at least by the rules of science. In short, belief is divorced from objective knowledge and can claim legitimacy only within its own private sphere.

Once the claim for divine judgment is dismissed in the secular court of adjudication, the matter sits, for science has largely defined what constitutes knowledge—objective truth—and more specifically, what is a *fact* as opposed to an *opinion*. In this sense, the dictionaries are correct: science is about true knowledge, systematic knowledge of a sort not casually found nor applied. However, the application of such standards to all aspects of human thinking must fail and a more integrative strategy must be sought for addressing matters beyond the particular confines of scientific inquiry. So instead of promissory notes based on expected scientific progress, we are better served by conceiving of science as a tool to help compose pictures of reality that must be coordinated with other belief systems. In other words, while scientific knowledge has become constitutive to the way we think of the world and ourselves, those understandings are tempered by personal experience and interpretation. When science is viewed with this larger view in mind, the objectivity of scientific pursuits couples with various forms of subjective judgments (both in determining social policy in a political context and on a personal level, one's search for existential or psychological understandings). In that synthesis, the place of science in Western societies becomes a means of reframing the humane questions that sponsor science, and even redirect current concerns over scientism to a more measured appreciation of science's accomplishments, current inquiries, and future promises.

Science and Its Values

The technical character of science obscures how science resides in a much larger social forum than the laboratory, or even the university or industry. For over five centuries, the received view of scientific investigation has highlighted how a methodology of discovery and verification that couples a uniquely rigorous empiricism to a critical rationality has evolved. Underlying these two components resides a commitment to scrupulous Baconian impartiality and Galilean neutrality (Lacey 1999). These values comprise the bedrock of science's ethics, and in this moral realm science frames its epistemological character. While methodologies of the laboratory and their products have captured the most interest, these are based on science's own moral code of open inquiry, pluralistic discourse, fallibility, and transparency (at least ideally). Science's epistemology, in fact, is value-laden, and some values, the epistemic ones—parsimony, coherence, predictability, and so on—have been shaped by a deeper set of values governing science's rationality. (For more on epistemic values, see Introduction, n. 1.) Indeed, though a more pragmatic orientation has replaced the positivist ideals, the ethical edifice still stands strong.

The ethics of science are inseparable from science's forms of knowledge. This thesis, a synthesis if you will, I have called a *moral epistemology* (Tauber 2005a). ("Moral" in the sense used here refers to value broadly understood. Good and bad are values, but so are objectivity and neutrality.) The most obvious expression of this moral-epistemological alliance is found in the discursive mores governing the open discussion of investigative findings, namely, the comparison of data, the free inquiry about interpretation, and the transparency of experimental reports. These parameters of discourse are based on the honesty of the participants; indeed, honesty is assumed as a simple epistemological requirement. Private experience cannot become a public fact until scrutinized by a community of observers. This might be achieved through direct experiment in multiple laboratories or, alternatively, through reports of investigations, which are then subject to review and criticism. In either case, the shared experience is crucial and confers the final criteria of objectivity, that is, diverse perspectives converging on a common assessment.

This social appraisal is, by definition, ethical inasmuch as the social is ethically constituted by particular rules, practices, and definitions. Simply casting science as a social activity confers a moral dimension to its characterization. And as an ethical system, science serves as a paragon of certain virtues society holds in high esteem: it is pluralistic and nondogmatic (i.e., accepting, detracting, as well as integrating of criticism as part of its very code [Popper 1945, 1963]). In this regard science is a bulwark of liberal, democratic society (Merton 1973). In respecting that science does indeed seek truth by such principles, we must be wary of confusing its ostensible and largely attained moral goals with the exceptional cases of dogmatic attitudes or fraudulent practices that threaten to subvert the ideal.

However, more insidious ethical dilemmas have developed over the past thirty years as a result of the growing intimacy between industry and academic research (Greenberg 2007). Conflict of interest between the apparent independence (read neutral and objective stance) of the investigator and the material advantages derived from rewards of corporate profit have raised disturbing questions about the correct relationship between private gain and public-supported research institutions and their faculty

(Kornberg 1995; Schwartz 1995). While hardly a novel relationship, the intimacies of laboratory investigators and their financial supporters have become more complex as private capital has increasingly supplemented government support for research. We will reexamine this matter in chapter 5. For now, suffice it to note that the ethics of research go far beyond the domain of honesty in generating and reporting investigative results. Scientific research can no longer be segregated into simple "pure" and "applied" categories, for virtually anything generated in the laboratory may promise financial gain to the investigator. With the growing alliance of industry, venture capital, and university research, meticulous contracts and offices of "technology transfer" which mediate the financial relationships between these various interests as patents, licenses, and profit-sharing, have introduced an entrepreneurial overlay on the pursuit of truth for its own sake. Indeed, the study of nature extends well into the "business" of scientific research, where commercial efforts complicate the innocence of exploring nature.

These financial concerns represent only an aspect of the larger ethical framework in which scientists conduct their work. Given the centrality of the ethics of how scientists govern themselves as truth seekers *and* how the moral dimensions of the scientific worldview reflect the interpretation of scientific knowledge in a framework oriented by human need, we would do well to consider science cast in a *moral* framework. Here, moral is considered broadly: the rules that define objectivity and differentiate knowledge from opinion; the mores of scientific conduct; the role of nonepistemic values in scientific inquiry and evaluation; the larger ethical context of scientific knowledge applications; and, finally, the personal dimensions that would integrate science's worldview with existential understandings and beliefs. In this widened moral context, the values *of* science and the values *using* science place investigative findings in the on-going construction of social and individual realities. Typically, philosophers of science see this exercise more narrowly, namely as placing facts within encompassing conceptual theories or models. I suggest that, before facts, before theory, the *values* of science command the character of knowledge, both in its production as well as its application.

On this view, ethics is understood as establishing and guiding the foundations of scientific practice *and* interpretation. Just as values guide the behavior of moral agents, so too do values guide the practice of scientists. The rules, having followed a winding historical path, effectively establish the groundwork by which scientists proceed in their particular projects. These are not just conventions, but rather instantiate the lessons learned of how best to accomplish the local tasks at hand. Accordingly, without the particular values of scientific practice and assessment, we cannot differentiate scientific conduct from other forms of inquiry; indeed, science has often suffered a blurring of those boundaries (Pickering 1994).

The public nature of scientific practice has raised new concerns regarding research fraud. That self-interest remains operative should surprise no one. After all, scientists naturally pursue their own best interests by promoting funding for their projects (e.g., Greenberg 1967, 171–209) and personal advancement generally is a constituent of professional life. However, some stretch the standards of ethical behavior, as do any proportion of humans arrayed on the ethical spectrum. Needless to say, slippage between the cup and the lips occurs often enough to conclude that scientists are hardly immune from human foibles. Nevertheless, such breaches of trust occasion self-righteous indignation and attacks on scientists' authority (Greenberg 2003). Furthermore, public mistrust arising from various celebrated cases of scientific misconduct has threatened the scientist's autonomy (e.g., Chubin 1990; Bulger, Heitman, and Reiser 1993). Although no reliable statistics exist, rising public awareness of research fraud threatens the very legitimacy of science (Lafollette 1996; Judson 2004).[7] These new concerns for accountability have challenged the very ethos of scientific practice. Society's ever-increasing investment in scientific pursuits has altered the relationship of esoteric knowledge and public access, with unpredictable consequences for both governing scientific conduct and the direction and speed of its future growth. Thus we see a vivid example of science's blurred boundaries in government action imposed on research institutions to ensure trustworthy research (Chubin 1990). And, not surprisingly, the standing of science as a model of ethical behavior puts its practitioners in a particularly hot moral spotlight.

While recent attention has focused on scientists breaking the ethical code, their normal conduct has contributed a consistent moral lesson to Western society. Indeed, science itself, with its distinctive social mission, open exchange, and rigorous standards of rationality, in large measure *defines* science. This moral posture presents science as an idealized venture towards truth and thus may be understood as representing a core human ideal toward which our society must aspire (Cohen 1974). Offering a model of both knowledge and ethics, science promotes a kind of moral activity that might appear unique to itself, but actually serves as a paragon of discourse in liberal societies.

The intersection of science and ethics is easily traced from the origins of Western moral thought in Hebrew and Greek sources, where theories of knowledge and theories of ethics were not only analogous but closely linked. This intimacy is evident throughout modernity, from Bacon to Kant and beyond. Science becomes a dominating influence in molding epistemology well beyond the laboratory, affecting all forms of knowledge production and adjudication of evidence (e.g., the judiciary). Further, notions concerning human nature and social structure derive from interpretations of scientific findings, which, in turn, influence the moral structure in which we regard our personal identities (e.g., as citizens). Scientific methods may provide a logic for moral discovery; scientific investigations offer "the data of ethics"; scientific achievements determine, in the broadest sense, the scope and limits of responsible moral choice; the ethics of scientific investigation serves as a model for societal behavior and truth seeking; and, ultimately, science poses for us the frontier of new problems and new circumstances for old problems.

The deeper lesson from what Hans Reichenbach called "scientific philosophy" (1951) is that the struggle to achieve a scientific worldview, with all of its problems concerning objectivity and realism, has provided us with rigorous criteria for discerning the limits of knowledge and, closely related, with appreciating the basis of logical choices in the ethical domain. Claims for how science might contribute a valuable core of objectivity to ethics have hardly been accepted in all quarters where the attempt to attribute a special truth ethic to science has itself been subject to reassessment. Those who see the fall of science from its domineering pedestal, where a naive positivism commands Truth and Reality, argue that science's posture relative to directing moral inquiry must be regarded with suspicion: how can science be neutral if scientists are self-evidently social creatures with political and psychological biases? While science still holds a pivotal place in Western societies, it has increasingly suffered assaults to its privileged standing. Indeed, why should science's governing epistemologies hold a privileged *moral* standing?

Since Reichenbach's claims advocating logical positivism, science studies have fueled the opinion that science is very much like any other social activity, and that the same general cultural rules that direct other complex cultural institutions also govern science (e.g., Pickering 1992). The moral theme then closely aligns with the revisionist epistemological picture, and thus the second moral dimension of science as a moral epistemology pertains to the values embedded in knowledge. I am not referring to how knowledge is valued, although this is highly relevant to this discussion, but rather how values are embedded in knowledge itself. This aspect of science's ethics is most evident in the normative sciences, that is, the life sciences, which evaluate functions as fulfilling certain teleological criteria (see chapter 5). Here, a normative spectrum is inseparable from scientific descriptions. The implications for the medical sciences are self-evident (Tauber 1999a; 2005a, chapter 1), but when the fact/value distinction is more broadly understood, we see that facts may lose their clear demarcations in the natural sciences as well. Instead of some positivist ideal of dispassionate, detached scrutiny, the interface of human need is imposed on the putative neutrality of scientific inquiry, and the complex interplay of facts and values issues a challenge to defining objectivity. After all, communal agreement underlies any notion of objectivity (e.g., Megill 1994), for knowledge is social, that is, knowledge is constituted by social practice (Pickering 1992; Schatzki, Knorr Cetina, and Savigny 2001) and embedded in social groups as a reservoir for use and identification (e.g., Latour 1987, 1988; Nelson 1990; Knorr Cetina 1999; Kusch 2002).

The definition of "social" focuses much of current debate about science's truth claims and the status of its knowledge (Hacking 1999; Kukla 2000), but beyond these particular issues, science as a human activity demands values to define itself. Some of these values derive most directly from social practices, others from a system of metaphysics, and yet others from particular legal, political, and moral philosophies. Recognizing this complexity has provided a heightened awareness of how science may be co-opted by politics and ideologies. Overtly (or subtly) influenced by its cultural milieu and moral environment, science never stood alone in an objective chamber insulated from the social pressures surrounding research and theory. Showing how science is so embedded does not diminish science's own accomplishments, or even its aspirations, so much as highlight how isolating science from what is typically regarded as its competitors distorts science's own character.

Instead of the various attempts to dissociate facts from values, which are seen as contaminating objectivity, we have come to understand that facts are facts because of the values that confer a factual status.[8] More, those values do not reside in some Platonic ideal domain from which their application confers the objectivity sought by investigators. Not only has the independent status of facts been radically reappraised, that is, facts become facts *because* of the values attached to them, their meaning and significance are determined by the context in which they are formed, appraised, and employed. In short, values are constitutive to facts, and, more to the point, *different* values determine how we understand facts in various contexts. Simply stated, facts cannot be separated from the values which embed them, and the understanding of science—from its institutional-political commitments to its various determinations of human comprehension of the world and human character—cannot be apprehended without some basic appreciation as to how values frame *everything*.

The notion of an insular "fact" belies how facts are so comingled with the values and theories in which they are embedded that to disentangle the relative roles of these supports becomes a highly convoluted, and sometimes irresolvable, endeavor. Thus the blurring of the fact/value dichotomy, both within the laboratory and outside, represents the overriding characteristic of postpositivist science studies. This position argues that a relaxation of the rigid fact/value dichotomy recognizes that science continually evolves diverse value judgments regarding its own practice that are never steadfast, but always changing in response to new demands and contexts. Chosen and developed, they hardly stand stable. No formal, final method exists to define fact/value relationships. When theory and fact conflict, sometimes one is given up, sometimes the other, and the choice as often as not is made "aesthetically," by adopting what appears to be the simplest, the most parsimonious, or elegant, or coherent—qualities which themselves are *values*. These are what Putnam calls "action-guiding" terms (1982), the vocabulary of justification, also historically conditioned and subject to the same debates concerning the conception of rationality. The attempt to restrict coherence and simplicity to predictive theories is self-refuting. The very logic required even to argue such a case depends on intellectual interests unrelated to prediction as such. In short, by dispelling the intellectual hubris of the scientific attitude, we are left with a more dynamic, albeit less formal, understanding.[9]

These concerns, seemingly restricted to philosophy of science, actually have enormous import. So much of what we generally understand as science, which is rightly considered a paragon of rationality and progress, we have associated with an objectivity that escapes the vicissitudes of prejudice and bias. But if, as a result of recent reassessments, we can no longer utterly separate facts from the values that support them, the common understandings of neutrality and objectivity teeter. The next four chapters describe the dispute arising from the crisis over objectivity—why the deconstruction of objectivity attained prominence, how that reevaluation has been represented and misrepresented, and where the critique elucidates the wavering line between scientific knowledge and its applications. We begin with the idealized view of objectivity from which all later criticism diverged, namely the philosophical claims of the nineteenth-century positivists.

2

Nineteenth-century Positivism

We have lost all reverence for the state. It is merely our boardinghouse. We have lost all reverence for the Church; it is also republican. . . . We have great contempt for the superstitions and nonsense which blinded the eyes of all foregoing generations. But we pay a great price for this freedom. The old faith is gone; the new loiters on its way. The world looks very bare and cold. We have lost our Hope, we have lost our spring.

Ralph Waldo Emerson, "The present age"

Beyond the appalling scientific illiteracy of the American public resides a profound ignorance about the nature of scientific institutions and the political infrastructure of research, not to speak of a lack of understanding of what constitutes contemporary scientific method, theory construction, and all the rest that goes into scientific practice. Instead, most hold a view of science that mimics the aspirations of a nineteenth-century ideal. That ideal was based on a characterization that presented scientists as a new priesthood in service to the pursuit of truth derived from radical objectivity. This picture of science was formally understood as a philosophy, *positivism*, whose ascendancy in the nineteenth century eventually defined scientific practice based on methods developed for the physical sciences. The life sciences soon followed, and, by the end of the century, intellectual debate swirled around how appropriately to apply the standards of the natural sciences to the human sciences (sociology, anthropology, and psychology [Smith 1997, 2007]). The debate expanded beyond the walls of the universities to include concerns uttered by humanists, theologians, and other skeptics as to what might restrict the dominance of science over other modes of knowing. And perhaps most importantly, all might well have wondered in what ways this scientific vision of nature defines reality (as already discussed). These issues defied simple responses then and now.

Positivism consists of four major precepts:

(1) nature might be observed without distortion of human cognition, which depends on a notion of objectivity that requires a radical separation of observer from observed, so that no subjective values are allowed to play in the gathering and analyzing of data;
(2) facts emerge from data, and those facts may be assembled into models and theories, which are then tested;
(3) reality is integrated, and scientific methods can be applied to study all phenomena—physical, organic, psychological, and social—by the same objective means;
(4) progress characterizes scientific pursuits, and faith in that progression promises evermore comprehensive laws of nature. At least, so it was thought.

Accordingly, from facts determined by objective methods, scientists derive hypotheses that are closely examined by experimentation. They then place these hypotheses in some ordered construct, which, in turn, is formalized in predictive theories more successful than previous ones. Several assumptions in this formulation require mentioning. The first is that the inductive scheme by which individual empirical observations are generalized "presupposes metaphysics." Alfred North Whitehead aptly referred to this basic presupposition, "an antecedent rationalism" (Whitehead 1925, 62). The method based on this assumption obviously "works," in the sense that such inductive reasoning has met with high success, but as David Hume noted with suitable skepticism, *why* it works is not logically self-apparent.

A second profound metaphysical assumption builds on the lingering Aristotelian notion of natural kinds and the "thing-hood" of nature's objects which science examines. These entities are assumed to exist as contained within a *simple location* of placement (Whitehead 1925, 69–70), which, in turn, depends on a particular understanding of the space-time continuum. Twentieth-century physics radically upturned a universe of discrete objects existing in fixed coordinates of space and time. This is important for our discussion because, with a simpler mechanistic philosophy of physics, "the real" is effectively localized and captured as objective entities. Such "things," waiting in nature for human discovery, Whitehead called the "Fallacy of Misplaced Concreteness" (1925, 72), by which he meant that the abstract descriptions of nature arising from modern science have paradoxically been conceived as concrete realities. In other words, what we might consider as things extracted (and ultimately abstracted) from nature are artificial constructs of our methods and interpretations. For Whitehead, a more precise description would acknowledge that "things" behaved more as "processes"—emerging, evolving, and, most importantly, only captured as "things" upon human measurement and abstraction. This picture of reality originated in the revolutionary findings of quantum mechanics, where the so-called "measurement effect" essentially froze reality upon human intervention (observation, measurement, assessment). (Paradoxically, a particle only exists in one place or another once measured, otherwise it may be "somewhere else." Only by *looking*, does the particle find its place![1]) Whitehead extrapolated the significance of such human interruptions to the macro-worlds of non-atomic physics and the life sciences, where he saw the discovery of entities as actually the construction of things, frozen in their peculiar fashion by the scientific methods of examination. He thus hoped mechanistic physics would yield to a science of process.

While Whitehead's "process philosophy" has yet to generate a direct influence on contemporary science, he (and others) did open the door to a new line of philosophical criticism. Positivist philosophy had asserted that investigations yielded facts, which in the everyday world of research meant that nature's objects and processes were independent of human interaction. With the quantum revolution, that position no longer could claim legitimacy.

Positivists held that our picture of reality appeared as if humans did not participate in the process of discovery, when in every sense, humans *made* facts, albeit from natural phenomena. Their fundamental conceit asserted that the objective data collected and facts sifted had expunged the human factor to reveal the world as it *really* exists. Accordingly, human intervention leaves nature essentially unperturbed, at least to the extent that objectivity yielded things as they were *in fact.* But quantum mechanics showed that this was not the case, and more generally, any intervention carried an entire set of interpretative problems. So the challenge presented by the most fundamental physics up-turned positivism's third critical assumption, the radical objectivity in which facts are conceived. Besides the older concern of how objectivity might be compromised by "subjective" values, the positivists now had to consider a science that fully acknowledged human presence and factored in human observation. The implications were difficult to over-estimate.

In many respects, the impact of quantum mechanics represents the critical turning point in positivism's fortunes, and to appreciate the significance of this reappraisal, let us briefly review the older positions. In the nineteenth century, positivists had effectively invoked critical distinctions between scientific facts and the values that threaten to contaminate them. In short, facts and values resided in split domains. This splitting of facts and values did not include the value of objectivity, which in its nineteenth-century form became the cardinal precept of the positivists. For them, objectivity radically replaced the personal report with one written in a neutral voice and a universal perspective, or, in other words, a report that might have been written by anyone given the particular setting and circumstances of the investigation. Thus, because true knowledge possessed no individualized perspective, a community of observers would warrant the findings. Agreement on the significance of a finding testified to the veracity of the facts under discussion, and then the significance and meaning of the facts might be discussed. In the end, a hypothesis, or even a theory, would emerge. Universal accessibility and a view from nowhere (i.e., independent of personal bias) became the key criteria of a new science.[2]

This move from the private sphere of experience to a communal universal had begun at the dawn of modern science, but in the mid-nineteenth century this ideal of truth became clearly enunciated as a scientific principle. The positivists' position comes from David Hume's famous eighteenth-century proclamation that one cannot infer an "ought" from an "is." This means simply that a moral case cannot be deduced from a natural fact. The critique is sometimes referred to as Hume's Law, which attacks the apparent rationality of various ethical or religious positions.[3] We can trace the later attempt to radically separate facts and values to Hume's original argument against the illogical deduction of religious belief from natural facts and morality from similar constructions derived from natural law or other systems of supposed rational basis (Putnam 2002).

He argued instead that ethics and religion are grounded in human emotions, needs, and caprices that are rationalized into religious dogma and moral justifications (Lindley 1986). Developing that issue takes us too far afield, but the salient point is that his philosophy supported the scientific aspiration of objectivity, that is, facts divorced from contaminating personal values. Much of nineteenth-century philosophy of science and the practice it guided was based on extrapolating Hume's cardinal insight; positivism successfully rejected subjectivity, which tainted the pursuit of "true knowledge." Indeed, the distinction of scientific facts and corrupting subjective values represents *the* crucial positivist distinction.

The status of facts in the modern scientific context dates to Francis Bacon's endeavor in the early seventeenth century to replace metaphysics with the concrete, the datum of experience. As Lorraine Daston notes, the word "fact" derives from Latin *facere*, "to do," and in the sixteenth century, the word still meant an action or deed. The critical Baconian distinction was that facts offered neither "consensus nor freedom from all bias, but simply freedom from theoretical bias" (1994, 45, 47). Daston maintains that facts became the focus of scientific discourse because they shifted attention away from the more contentious wrangling over rival theories, thus a social norm fashioned scientific practice:

> Since the academicians believed that partiality to one's own theories and opinions was the apple of discord rolled in their midst, the kind of impartiality they sought was impartiality to theory. . . . Therefore, the purportedly theory-free Baconian facts suited their purposes perfectly, despite their other obvious disadvantages. Thus did objectivity come to be about the impartial examination of neutral facts. (Daston 1994, 57)

The status of objectivity depends on multiple components fitting together. The first concerns the status of the observing agent. The invention of classical perspective in painting during the fifteenth century serves as a ready metaphor for the birth of modern science inasmuch as a self-conscious position is assumed to survey the world (Fox Keller 1994). Take a singular position, hold it, and then report what is observed, objectively. And just as painting assumed its particular styles of transmission, so did the language of the newly emerging objectified science. The scientific rhetoric thus adopted a neutral language, which reflected the self-conscious separation of the observing subject from the object of study to describe an objectified world, an established authorial authority, and, finally, a "cleansing [of] the lens of perception"

(Gergen 1994). Each rhetorical device, again, those being a detached observer and a neutral language, contributed to generate a radical sense of objectivity.[4]

In one sense, the idea of the detached observer was the first step in accurately assessing nature, but an even more radical rupture of subject and object was required. The task of modern science was first to standardize observation and then to eradicate the observer altogether in the quest of a complete elimination of the subjective dimension. By focusing on experimental procedures, Robert Boyle effectively propagated a shared research program, which generated a "multiplication of the witnessing experience" (Shapin and Schaffer, 1985, 488).[5] With public demonstrations and the enlistment of other scientists, Boyle's rhetoric of reports emphasized the observed facts and described experimental procedures in great detail (virtual witnessing). The singular subjective observation was thus cowitnessed and translated into a shared public objectivity through the machine's results. The disjunction of subject from observation was hardly complete, however, and for objectivity to assume its current meaning of being "a-perspectival," extensive rhetorical refinement and the development of statistical analyses in the nineteenth century were required. Standardized equipment and techniques universalized scientific practice so that the first person report could be replaced by a universal anonymous one. The scientist assumed this voice and became an authority of how to achieve an objectivity that would leave the human only as a machine among machines. This was the positivist ideal, and this new persona carried profound implications and many denials.

Constructed in opposition to the romantic era view of the world, which privileged the individual's perspective and subjective experience, positivism denied any *cognitive* value to value judgments. Personal experience, positivists maintained, cannot be extrapolated into a scientific description. "Noble," "good," "evil," or "beautiful" are qualities of men or events, and while such adjectives may be applied to nature, in doing so, a projection of human sentiment is assigned to the phenomenon. In direct reaction against the romantics, positivists sought instead to radically objectify nature, banishing any and all human prejudice from scientific judgment. The total separation of observer from the object of observation—an epistemological ideal—reinforced the positivist disavowal of value as part of the process of observation. One might interpret, but such evaluative judgments had no scientific (i.e., objective) standing.

The romantics deeply understood (and resisted) the hegemony of the ascendant positivism, and they placed important caveats on the positivist approach to nature on both epistemological and metaphysical grounds. From their perspective, each inviolate observer held a privileged vantage, and they jealously protected this vision (Tauber 2001). Simply put, where the romantics privileged human interpretation (exemplified by artistic imagination), the positivists championed mechanical objectivity (data derived from instruments, e.g., thermometer, voltmeter [Daston and Galison 2007]). A common understanding dating to the nineteenth century portrays the scientist as vanishing, absorbed by her machines. As a simple reporter of her instruments, the subjective element is supposedly eliminated. But if one steps back from the persona of the scientist as a social entity and attempts to portray her as subsumed beneath the epistemological demands of the view from nowhere, a "paradox of scientific subjectivity" emerges (Fox Keller 1994). This refers to the ostensible goal of a completely detached observer, one independent of subjective foibles and prejudices, whose conclusions come from "somewhere else."

For positivist science, facts (of a certain kind) were to reign supreme. The argument between positivists and their critics, a debate revolving around the standing of facts, has framed philosophy of science debates into our own era. To further understand that history, we begin with the first clear separation of these vying conceptions of science.

The Argument

At the end of the eighteenth century, Johanne Wolfgang von Goethe developed a sophisticated philosophy of science that in many respects served both the later positivists and their detractors. Goethe, rejecting the allure of a radically objective science, appreciated that facts do not reside independent of a theory or hypothesis that must support them. His precept that "everything factual is already theory" (Goethe 1998, 77) was offered as a warning about the complexity of objective knowledge. The synthetic project of building a worldview proceeds dialectically, but from very different tenets: Facts become facts because of their supporting theory, which orders the observed phenomenon and conceptually defines their meaning. Reciprocally, the facts support the theory, which is a process that continues with integrating that scientific picture within the broader and less obvious intellectual and cultural contexts in which the larger conceptual apparatus is situated. Thus he argued that facts, as independent products of sensory experience, are *always* processed—interpreted, and then put into some overarching hypothesis or theory. In short, observations assume their meanings within a particular context, for facts are not just products of sensation or measurement as the positivists averred; rather, they reside within a conceptual framework, which places the fact into an intelligible picture of the world.[6]

Of course, Goethe understood the potential danger of subjective contamination of scientific observation, and, more to the point, the tenuous grounds of any objective fact that relied in any way on interpretation. The concept of interpretation stretches from inference to direct observation, for any perception must ultimately be processed to fit into a larger picture of nature and must cohere with previous experience. Moreover, in recognizing the claims of positivism, Goethe countered that the place of the observer in scientific discovery could not be completely omitted. Indeed, he embraced this faculty of judgment, broadly construed, as both the source of creative insight as well as a regulative faculty of great importance to the scientific venture (Tauber 1993). (Goethe's position would be developed in a *postpositivist* challenge by Polanyi, which is a topic of the next chapter.)

So, well before the positivists formally espoused their own agenda, Goethe clearly recognized the complex question of scientific identity between the detached observer, supposedly divorced from theoretical presuppositions, and the creative investigator: "my thinking is not separate from objects; that the elements of the object, the perceptions of the object, flow into my thinking and are fully permeated by it; that my perception itself is a thinking, and my thinking a perception" (Goethe 1823/1988, 39). This realization of a confluence between subject and object was later formalized and developed in twentieth century phenomenological philosophy, where the "gaze" is the privileged vehicle of the subject's relation to the world; consciousness and meaning depend quite literally on how we see things (Husserl 1935/1970).[7] The scientist must still endeavor ideally to objectify, but as Goethe also recognized, the integrating creative insight resided within a more complex faculty:

> This experimental reality, which is the *only* reality we live immediately (as opposed to scientific "reality," which is abstract and grasped intellectually rather than experimentally), is thus fundamentally subjective in nature. The objects that surround us function less "as they are" than "as they

> mean," and objects only mean *for* someone. . . . To see implies seeing meaningfully. (Morrissey 1988, xx; emphasis in original)

The inextricability of subject and object contradicts the ideal of the scientist as independent from the world—the austere observer, collector of data uncontaminated by projected personal prejudice.

How then do the crucial and variable elements of creative intuition, deduction, and assembly of disparate information create "objective" reality? Much of our understanding rests on a different, "non-scientific" intelligence where "events are not counted but weighed, and past events not explained but interpreted" (Heisenberg, 1979, 68). Accordingly, whether others have castigated or praised Goethe for his scientific philosophy, his argument ultimately reduced to the legitimacy of a holistic faculty that would seek an exhaustively comprehensive study of his subject. Despite his objective methods (and his rejection of Schelling's projection of "mind" into nature), an aesthetic sensibility guided Goethe's studies, which left him with an unresolved conflict: To what extent, as scientist, was he allowed to vent the power of his artistic intuitions? How might he freely acknowledge the legitimacy of aesthetic judgment in scientific discourse, and what aesthetic principle would he find useful?[8]

Goethe's holistic attitude, born in aesthetics, enjoyed strong support throughout the nineteenth century. For instance, Benedetto Croce (1902/1972), similar to Goethe, would call art and poetry forms of cognition and saw the aesthetic as discerning diversity within an encompassing unity. But they stood on one side of a deep fault line between science and the humanities. As the positivist and reductive strains of scientific inquiry gained momentum, the more general admonition to integrate the widest scope of experience was lost. While scholarly debate has extensively considered Goethe's scientific character (e.g., Amrine, Zucker, and Wheeler 1987; Bortoft 1996), he must, on my view, be regarded not as a "poet scientist," but rather as a "holistic scientist." To label him as a poet scientist inappropriately imposes our own divided sensibility of a Two Cultures world, when he regarded a unified nature with a unified mind integrating poetic and scientific sensibilities. The poet and scientist necessarily view the same object as differently refracted experiences, but for Goethe, the experience of the object must ultimately be integrated by an arbitrating observer. Through a synthesis of scientific reason and aesthetic judgment, he purportedly achieved unification of disparate ways of knowing, where "science and poetry . . . when properly employed [were regarded] as parallel and complementary ways of seeing" (Abrams, 1953, 308–9). This theme was pursued in diverse directions throughout the romantic era (McFarland 1969; Cosslett 1982) and inherited by the mid-nineteenth-century positivists, who put their own characteristic stamp on this philosophy in seeking a radical separation of objective and subjective modes of thought.[9] So, from their perspective, by labeling Goethe a poet scientist, he suffers the stigma of subjectivity.

Purging subjectivity from science constituted the major reorientation of the postromantic period. To view the world objectively is to remove the subjective ego from the encounter. That is, the scientist as the knowing subject must divorce himself from projecting bias and subjectivity onto his inquiry. Indeed, the nineteenth century left behind the multidimensional (aesthetic, historical, speculative) approach to pursue a new method, one that would seek a unified reason in ways Goethe would have disallowed. Positivism's aspiration for the unification of knowledge and a universal reason to pursue it took part of Goethe's scientific agenda and narrowed it to a single logic.

But the wheel again turned, and during the later decades of the twentieth century, science studies, dissatisfied with a positivism that sought to radically divorce the scientist from the object of inquiry, rediscovered how the context of study represents a crucial factor of scientific pursuit. These themes will be further explored later. Suffice it to note that some contemporary science studies interpretations do not radically differ from Goethe's own conclusions, nor do they deliver us from the imbroglios he so clearly understood. His holistic project has been fractured, and, despite the extraordinary accomplishments of the approach he opposed, the philosophical need to pursue a unified reason remains.

The positivists, of course, rejected Goethean holism because it was so tinged with subjective elements. A world built from their principles would appear essentially the same to all viewers because facts for them have *independent* standing and universal accessibility, so that irrespective of individual observations, facts constitute a picture of reality. From this orientation, the independence of the known fact rests on its *correspondence* to a reality that any objective observer might know. This assumes both a universal perspective—a view from nowhere—and a correspondence theory of truth.

Let us begin to probe the positivists' presuppositions. In regard to the relationship of fact and value, it seems obvious that we cannot easily divide these between two domains that have no overlap. Even the positivist standards applied to natural science represent values, historically arrived at and chosen in everyday practice (Putnam 1990).[10] Indeed, facts are facts because of the values assigned to them. But the positivist maintains that a natural fact reflects a natural reality, and the objectivist values employed by scientists are suitable for ascertaining such bits of reality. Philosophically speaking, their position rests upon the "correspondence theory of truth" (Lynch 2001), which states simply that our cognitive functions present the world directly to us. Common sense holds that we have perceptions and derivative language and symbols which depict that reality *as it is*. So, on that basis, facts are simply the currency scientists employ to know nature, and we can be confident of their worth because facts are "bits" of that reality. In sum, facts *correspond* to nature's reality in the positivist mind-set.

The alternate view is perhaps counterintuitive at first, or so obvious that it hardly seems possible that furious arguments have ensued on its claim: the world clearly exists, but the reality we perceive can only be *real* as *we recognize* it. A reality greater than the one we have so far perceived seems a reasonable inference, but that is not the question at hand. The issue is that reality is that which we *know*. Our brains have certain knowing capabilities that have evolved to allow successful navigation of the world. We are highly successful and we can control many aspects of nature, which proves the effectiveness of our capacities and confirms the confidence that we know the real. But human cognition is distinct to our species and thus the reality we know, the human intercourse with that reality, has a different character from that of my dog or the fly buzzing around my head. All three of us live in the same world, but our respective perceptions of what *is* are quite distinct, so our respective realities also differ. Indeed, even human reality changes in history. Before the telescope or microscope were invented, reality appeared quite different from the one we now appreciate, and with the relativity and quantum revolutions, the very notions of time and space have assumed a radically different configuration from that described by Newton. We now may have a more complete understanding of the real, but in what sense can we claim to know the real in any *final* sense? We are limited by our collective mind, and as powerful as it undoubtedly is, limits are at play. In other words, the *human mind and nature together comprise reality*. This latter formulation has several philosophical expressions, but for now, we will discuss the epistemology of this mind-dependent world as a kind of *constructivism*.

Constructivism, at least in the context of our discussion about science, refers to the process of *constructing* the context in which facts are placed, that is, in a model or theory. For the positivist (who holds to the metaphysical realism discussed above), facts are real, and we just have to be clever enough to fit them into the puzzle, and thereby understand their proper significance and meaning. For the "anti-realist" critics, facts are constructed to some degree, and the debate among them is how much. Later chapters are devoted to this issue, so here we only note that while *context* has proven a highly plastic concept, we may distill the debate into a major dispute about whether facts fit into a mosaic scientists *discover* (and are therefore "real"), as opposed to placing facts into a framework which has been constructed from various elements. Those "extracurricular" pieces are not just the furniture found in the natural world, but also include powerful cognitive components derived from culture, language, and history.

With this background, we come back to the fact/value dichotomy. The question devolves to an argument about the values employed to build the scientific edifice, namely, to what extent epistemic (pertaining to knowledge) and nonepistemic (social, cultural, linguistic, emotional, etc.) values play in portraying the reality science offers. Contemporary science studies have concluded that, as persistent as the positivists might have been in attempting to stamp out subjective influences, they only succeeded in making them seem disreputable (Daston and Galison 2007).

The story is further complicated by the stark distinction between theory and observation made by nineteenth-century positivists. They too realized that placing facts (observations) within a larger construct, usually a model or theory, introduces a bias, even a circularity: the interpretation and the fact support each other: the theory places the fact within the larger construct, and the construct is supported by the fact. In other words, the fact is a fact because it fulfills the criteria that permits its placement in the theory that requires such facts, and the theory, in turn, feeds itself with facts it has construed as factual. Considering their passion to excise biased interpretation, such potentially subjective commitments or prejudice remained troubling to nineteenth-century positivists. They had highlighted this problem as crucial to their own undertaking because they argued that, to have objectivity, observations must be freed of all bias, and they were self-consciously aware that bias extended to every aspect of the scientist's observation, from data retrieval to interpretation.

Accordingly, psychological projection, self-interest, and, most abstractly, commitments to a particular hypothesis or theory, might contaminate the sanctity of scientific facts and their proper interpretation.

Thus distinguishing facts from nonscientific values provided a crucial element of the positivist program. The attempt to control for subjective bias explains, at least in part, later efforts to divide the criteria and logic of investigative "discovery" from those of "verification."[11] By differentiating the processes, observation putatively moved further away from interpretation, allowing that (assuming that) each required different modes of objectivity. In short, only if observations were independent of theories could they serve as evidentiary warrants of a theory's adequacy, and that task required the control of contaminating prejudice. Contemporary science studies accept these kinds of distinctions as serving an ideal, but meticulous observation of practice (which I will discuss in chapters 4 and 5) has shown them as only idealizations and working standards.

The Knowing Agent in Dispute

There is no escape from the constraints of an observer fixed by her individual perspective, contextualized in some observational setting and committed to processing information through some interpretative schema. Such an observer cannot adhere to a rigid identification of facts based on an idealized separation of the knower and the known. Various kinds of values knit the factual world together into a more or less coherent worldview (Tauber 2001). The strategies vary. The social, linguistic, and political effects often lie hidden. The ideological influences invariably remain subtle. The commitments to obscure conceptual structures usually rest dormant. But in the end, each of these extracurricular influences is in force, and the struggle for objectivity becomes just that: a struggle. In the ongoing negotiation with the natural world, on the one hand, and the community of scientists and their partners, on the other hand, scientific truth emerges . . . not in any final form, but only as a tentative statement of what constitutes the real.

These broad conclusions, of course, are the product of our own era. In the nineteenth century, despite some appreciation of the caveats described above, the radical separation of the observing/knowing subject and her object of scrutiny served as the single most important characteristic of positivist epistemology. Because of this understanding, positivists claimed that science should rest on a foundation of neutral and dispassionate observation: the more careful the design of the experimental conditions, the more precise the characterization of phenomena, the more likely the diminution of subjective contaminants. Thus the strict positivist confined herself to phenomena and their ascertainable relationships through a vigorous mechanical objectivity. That model, developed most assuredly in the physical sciences, was quickly translated to the biological and social sciences, which stretched the positivist standards to accommodate a different epistemological orientation, one framed by normative standards (Tauber 2005a).

This general approach was not limited to the study of the natural world, for by the 1850s, positivism came to be understood as a philosophical belief which held that the methods of natural science offer the only viable way of thinking correctly about human affairs. Human subjectivity now resided under a new lens of inquiry, and with this mandate, the human sciences emerged (a matter discussed below [Smith 1997, 2007]). Accordingly, empiricism, processed with a self-conscious fear of subjective contamination, served as the basis of all knowledge. Facts, the products of sensory experience and, by extrapolation, the data derived from machines and instruments built as extensions of perceptive faculties, were presented as self-sufficient entities; "hypothesis" was defined as the expectation of observing facts of a certain kind under certain conditions; and a scientific "law" was understood as the proposition that under certain conditions of a certain kind, facts of a certain kind were uniformly observable. Any hypothesis or law that scientists could not define in terms such as these would be written off as "pseudo-hypothesis" or "pseudo-law." A newly construed attitude would regulate the use of such terms as *knowledge*, *science*, *cognition*, and *information*. Thus the sciences separated themselves from older traditions of inquiry.

In summary, nineteenth-century positivist proponents, who regarded scientific growth as synonymous with modernity and progress, embraced a method whose values have not only bequeathed an increasing capacity to control nature and raise human standards of living, but also provided a powerful, albeit particular, means for understanding the world and human nature. Dogging that promotion, a critical chorus saw science as distorting human life and imperialistically dominating other modes of experience. In its endeavor to seek some final truth (defined by its parochial methods) and to embrace its own mode of rational discourse, critics with this perspective regarded science as overwhelming other kinds of knowledge, most notably those oriented by humanistic values and concerns. On this view, the relentless assertion of positivist values—the scientism arising from the projection of such values beyond the domain of the natural sciences—provided narrow and distorted views of human nature and society. Some of the contentious issues in the Science Wars (see chapter 4) originated in these nineteenth-century debates about the character of reason and what counts as rational.

The positivism that arose in reaction to romanticism framed a group of questions derived from the radical break between subject and object that the positivist averred as necessary for scientific inquiry. From such objective knowledge, the romanticist asked, in turn, how scientific knowledge becomes *personally* meaningful. Or as Schiller asked, "How are we to restore the unity of human nature" in a disenchanted world? (1801/1993, 121). To illustrate this issue, consider Henry David Thoreau, who offered a philosophy actively engaged in critical dialogue with positivism. Although we have a different conception of science than the one he combated in the mid-nineteenth century, the basic issues remain the same.

Thoreau powerfully depicted how individual judgment in all of its multitudes of expressions holds nature and the individual together. For him, beyond the requirements of scientific scrutiny, which he actively pursued, creative effort places facts, both individualized and discrete, into some larger structure. For Thoreau, analysis fundamentally was a *self-conscious* interpretative act that synthesizes the perceptions of the world into a construct that confers meaning. Judgment mediates the action by which the individual places the data within a theoretical framework or, alternatively, incorporates their observations into an aesthetic, moral, or spiritual framework that confers personal meaning to the observer. In each case, the individual makes a self-conscious, interpretative, creative judgment. And judgment is value-laden and thus undeniably individualized (Tauber 2001).

Thoreau penned his most imaginative writings, the result of his "experiments in living," during the pivotal moment of romanticism's ebb and positivism's rise. Indeed, during the 1850s he expertly practiced "natural history" both in his own amateur pursuits and as a specimen collector for the newly arrived Harvard professor of zoology, Louis Aggasiz (Tauber 2001, 122–24). (That faculty appointment signaled the professionalization of biology in the United States, which included a newly conceived organization of the academy between the sciences and the humanities.) Thoreau's scientific endeavors involved careful measurement and observation in many different venues. He certainly comprehended how to apply systematic thinking to a problem, whether formulated as a record of the first appearance of flowers or as the technical challenge of determining the best mixture of clay and graphite to make a better pencil. In other words, Thoreau knew what it meant to engage in the science of his day, and reasonably followed the standards of geology, taxonomy, ornithology, entomology, botany, and ichthyology in the naturalist tradition (Walls 1995).[12]

However, from another vantage, Thoreau saw the stakes at risk in the ascendancy of a new scientism that accompanied the rising tide of positivism, which began to sweep the scientific community of the 1850s. Like other romantics, when he repeatedly asked what the relationship between the object of inquiry—the natural world—and the method of study and reporting was, the answer appeared to displace the individual from the inquiry. For him, objectivity obstructed his individualized vision of what a description of nature must attain. He acknowledged the importance of scientific inquiry; he denied that this was the end of his studies:

> I fear that the character of my knowledge from year to year becoming more distinct & scientific—That in exchange for views as wide as heaven's cope I am being narrowed down to the field of the microscope—I see details not wholes nor the shadow of the whole. I count some parts, & say "I know." (Thoreau 1990, 380)

His epistemological endeavor contains multiple layers and as judged by positivist standards, would meet with only varying success—swinging between detailed observation of all forms of nature to a distinctive prose poetry written in the genre of nature vignettes (Tauber 2001, 81–83; 91–92). Thus Thoreau's natural history is a complex array of several modes of knowing and an overlapping of several kinds of writing. He invoked different rationalities. To appreciate his achievements, we must be sensitive to the role each played in mediating his experience.

Of course, the positivists made no allowance for this more global experience, while the romantics made this challenge their chief concern. For the latter, the study of nature offered a personal comprehension of the world, a picture of reality that suggests insight into, and thereby an orientation of, humans in nature. The critical step moves the observer (the scientist) from *outside* the "picture" to a subject *within* the picture. And thus we have the basic conflict between two ways of looking *at* the world and being *in* it.

Two Cultures

The term *scientist* was coined by a British scientist and philosopher of science, William Whewell. In 1840, writing in the introduction to his *Philosophy of the Inductive Sciences*, Whewell commented, "We need very much a name to describe a cultivator of science in general. I should incline to call him a *Scientist*" (cxiii). The definition itself is not noteworthy, but the late date of its birth is. After all, the word "science" is ancient. The Latin *scientia* means "knowledge" as opposed to *sapientia*, wisdom. In other words, *scientia* is knowledge of, or cognition about, the world, as opposed to the more self-reflexive domain of wisdom. And *sciens*, "knowing," originally meant "to separate one thing from another, to distinguish," which also points to analysis of a particular kind. Certainly, this etymology closely adheres to what we broadly understand science to seek.

In short, the word *science* has an ancient etymology, but "scientist" is distinctly modern. Indeed, Charles Darwin, who wrote during the same period as Whewell, referred to himself as a "natural philosopher." Darwin carefully composed his language, and as a gentleman, he had good reason to prefer the older designation. The term *philosophical* was not explicitly defined but generally stood as the study of the natural world that included the search for laws in biology, a dissatisfaction with teleological arguments, a certain speculative or intuitive attitude in method (especially rampant amongst the *Naturphilosophen*), and a general commitment to an idealist approach (Rehbock 1983, 3–11). In addition, "scientist" was too easily associated with commercial overtones of technical applications, and thus the designation carried a pejorative connotation of someone who was inclined to look for the economic benefits of discoveries in contrast to the pristine search for true knowledge. Not until the end of the nineteenth century would the term scientist assume more neutrality.

Thus, until the mid-nineteenth century, science was a category of philosophy. The examination of the natural world was part of what philosophers did. Only as the methods of scientific inquiry became increasingly technical, and a new professionalism took hold in its various disciplines did a scientist emerge as someone different from a philosopher. If one examines the Western intellectual world as late as the American Civil War, the educated classes were comfortably conversant with the latest scientific findings, and many pursued what we would call "amateur" science (Tauber 2001, 121–31). Chemistry and physics began to separate a bit earlier, but certainly natural history remained the province of a wide audience. I am not referring to its popular mode: gentlemen would go to natural history meetings well into the 1850s and 1860s without any professional encumbrances to their full participation. In short, until about 150 years ago, most scientists and philosophers shared the same intellectual bed (Postlethwaite 1987).

However, advances in scientific techniques and methods of study required specialization. The techniques developed in the nineteenth century reflected a growing sophistication in terms of both material investigations and the mathematics supporting them. The field of "biology" was invented as its own discipline in the first decade of the nineteenth century, and by the 1820s, Claude Bernard (1865/1927; Holmes 1974) and other physiologists were reducing organic processes to physics and chemistry (Galaty 1974; Lenoir 1989; Kremer 1990). Concurrently, physics and chemistry were employing new mathematics, primarily statistical in nature, which by the 1870s created statistical mechanics and all that it spawned. Focused attention to technical knowledge became a prerequisite for active participation, and this demanded specialized training. Eventually this professional narrowing led to academic and professional segregation in Western Europe and in the United States (Knight 1986; Bruce 1987). For instance, in Prussia, the number of students enrolled in universities quadrupled during the nineteenth century, while the number of those studying newly formalized sciences increased fiftyfold (Proctor 1991, 76–77). By the 1870s, science was divided into various natural and social sciences, each of which assumed a high degree of technical competence and cognitive training (Smith 1997).

The fruits of that labor resulted in new industries derived from scientific findings and their successful application to material culture. Since the Renaissance, science had been sold as a shrewd investment: develop scientific inquiry, and the discoveries will be converted into economic, military, and social power. Indeed, the overture has been true to its promise; few can dispute that the triumphs of technology are inseparably linked to the success of the underlying science. But technology is *not* science; the two are distinct. Technology builds materially on scientific insight and much else, while science seeks to discover the character of nature and is thus part of the philosophy of knowledge. On this view, technology is the *application* of knowledge for material innovation, while science underlies such engineering. But the close identification of science and technology often blurs this distinction. I mention these differences here to emphasize that science has been too often associated with its product, as opposed to its deeper commitments to philosophical inquiry, albeit of a special kind.

More importantly, the intellectual discipline of each domain drifted apart. The hermeneutical methods used in the humanities, writ large, have their own standing. But the interpretations applied to human creativity are not suitable for the study of nature under the present scientific paradigm. The various objects of investigation have evolved different approaches and different truth criteria. Those who would separate science and the humanities would do so primarily on these methodological divergences. Indeed, the separation is rooted in the nineteenth-century modification of positivism to a more radical format and its extension to the human sciences, one that increasingly sought to describe the world *objectively* (Simon 1963; Kolakowski 1968; Daston and Galison 2007).

Consequently, nineteenth-century positivism provided a philosophy for the sciences to claim a unique intellectual and academic territory. Those borders were jealously guarded and broke a long-standing arrangement. The humanistic-scientific alliance, rooted in the Renaissance and matured during the early modern period, remained integrated through the Enlightenment and split only during the mid-nineteenth century. With the professionalization of the scientist and his segregation within the laboratory, those who remained created their own domicile, the humanities.

Not surprisingly, given the segregation of the scientist, *humanism* (referring to the rediscovery of the classical tradition in the medieval period,) was coined (like the word *scientist*) in the nineteenth century. Humanists were originally concerned with a general education that spans the classics to modern science. But humanists came to be associated with the broader liberal agenda: freedom of thought, tolerance, revision and correction of opinion, open communication, and a self-critical attitude—all in the employ of furthering a humanist understanding of the world. That interpretation was founded on a human-based understanding (human reason as opposed to divine revelation) and human-centered inquiry (celebrating the autonomy of the moral agent and knower). Of course this hardly exhausts the humanist agenda, which would in the course of its development include the primacy of personal judgment, individual-derived freedom (against any form of authoritarianism), and an integration of varied forms of knowledge. Note that, rather than serving as a moat, these underlying precepts and the values supporting them bridge the central concerns of the humanities and the sciences into a powerful alliance. Accordingly, the scientific worldview could make its claims based on a long history of coupling its particular concerns to this larger program. After all, science originated as a contributing member of the philosophy faculty, and on this broad view, science is part of a historical development of humanistic thought.

Although science follows a naturalistic philosophy, its empiricism is based on a rationality that has deeper roots in natural philosophy. Scientific epistemology emerged directly from natural philosophy, which, in turn, was part of a comprehensive intellectual orientation: ruthless self-criticism leaves the frame of reference always in doubt; the historical record reveals fallibility; the place of objective knowledge, as opposed to subjective opinion, is tested and contested; and when opinion is held, it is open to revision through free argument. Science is sustained, indeed instantiated, by a self-critical philosophy, tested against the investigations of nature. These are the deepest values of science and the underlying philosophy guiding its methods and defining its aims. Nature devoid of human value and human caprices demanded honest answers to starkly posed questions. In short, although the sciences and the humanities pursue different objects of inquiry, they support each other in common purpose and share the same self-critical attitude. And beyond this kinship, we find other aspects that link them.

The rise of science helped pave the way for secularism's triumph and the ascendancy of liberal political societies. As discussed in chapter 1, the values at the foundations of scientific inquiry are often at odds with those of religionists. In particular, scientists instantiated a rationality that had become a tool for open-ended inquiry, with a mode of truth-seeking that distinguishes itself by refusing to be guided by or serve any predetermined goal; fallibility is assumed; objectivity is sought. Despite deep fractures in this idealized view, these epistemic values are fundamental to the success of science, and concomitantly, the rise of modernity. (The obvious contrast is argument directed by a religious faith that is constrained a priori by presuppositions deemed immune in advance to questioning.)

The conflict is clearly illustrated by the reception of Darwin's *On the Origin of Species*, published in 1859, which sparked a crisis over religious belief and metaphysics based on the divine. This turbulent theatre of contention drew distinct battle lines between various kinds of religionists and secularists. In the United States stalwart promoters of secularism (like Robert Ingersoll and Cornell's founding president, Andrew Dickson White [Feldman 2005]) denounced religion as an offense against science. Darwin's prescient early journal musings (July 1, 1838) soon became commonplace sentiments: "Origin of man now proved.—Metaphysic must flourish.—He who understands baboon <will> would do more toward metaphysics than Locke." (1987, 84e, p. 539). And even before *On the Origin of Species*, Whewell could assert with arrogant confidence, "Man is the interpreter of Nature, Science the right interpretation" (1840, xvii). He was undoubtedly echoing the opening inspirational (and wishful) aphorism of Bacon's *Novum Organon*, which three hundred years later had become a confident summation of science's actual achievement, or at least so Whewell thought. And by that assertion, he meant specifically the findings obtained by a radical objectivity guided by positivist principles of inquiry. (Cosslett [1984] offers a rich compendium of the nineteenth-century debate.)

Thus the drama was not limited to evolutionary science per se, that is, Darwin's theory of common descent, but about the metaphysics in which Darwinism functioned. The Darwinians argued that science's understanding of the universe and our place in it may or may not include a divine presence; put another way, God is besides the point. More than just rejecting religious doctrine, Darwinism asserted its own metaphysical picture in contrast to it, namely, a stark, materialistic universe with no *telos*. Such a view leaves to humans the chore of defining significance and meaning within a *human* construct. Imposing a secondary layer of divine interpretation upon those findings does not warrant conflating two ways of knowing. Each has its place and, therein, its authority.

How to achieve pluralistic balance lies at the base of the conflict between secularism and religious ideology. Whatever triumph science might claim, the various alternatives to knowing have hardly lost their appeal. Indeed, the pronouncement of science's promise is only one of many chapters of dispute that characterize the battle between science and religion as competing worldviews. From Galileo to current debate about intelligent design, the character of truth remains hotly contested.

If science were regarded simply as a tool for technological advancement or mastery of nature, the debate would have been quelled, but all understood that much more was at stake.

From divergent positions intractable argument ensued. After all, the answers science provided were hardly neutral inasmuch as the secularists regarded investigative findings with one set of lenses, while the religionists peered through another. Indeed, both parties violated the borders as they sought to bolster their respective programs. Theoretically, a strictly neutral science would posture itself towards neither camp, but given its historical and cultural affinity with the humanist tradition, science became a powerful instrument of humanist philosophy. Moreover, since neutrality was never a viable option, science found itself caught in the crossfire of an ideological war that has been waged for over five centuries. And no wonder, for no less than the "Truth" was at stake.

Nineteenth-century secularization signaled God's further retreat from the everyday world of common experience and activities and also highlighted a major realignment of social hierarchies and the rationale for new political structures. Science partook in this social revolution in at least three ways: (1) the technology based on scientific discoveries revolutionized the material culture, revealing mysterious forces and events as natural and thereby open to human understanding; (2) this naturalized worldview made divine intervention increasingly peripheral to human understanding; and (3) the logic and standards of knowledge as applied to the natural world were extended to the social and psychological domains of human experience, thereby rationalizing a redistribution of power and authority from monarchial and ecclesiastical centers to liberal institutions. These developments revised God's status, and as God's place in the universe shifted, so did humankind's. Here, the convergence of other cultural forces combined in the eventual triumph of secularism: the realignment of authority, the autonomy of the individual, the claims for individuality, the rise of free agency. (And, of course, strong arguments have been made as to how post-Reformation Protestantism also contributed to the rise of modern scientific epistemology [Harrison 1998].) The success of the positivist program furthered the cause by demonstrating how a more rigorous objectivity, which had replaced intimacies of the heart with a different logic and a different understanding of the world could master nature. So, as the objective eye achieved primacy, science increasingly became the adjudicator of true knowledge.

The pervasive *philosophy* of science is inextricable from the political shift resulting from the rise of secularism. In a cascade, (1) the empiricist measures rational discourse against a natural object that "speaks" back to the individual observer; (2) private reports assume a new position as independent sources of knowledge; (3) autonomous reports require public confirmation, and thus objectivity attains a new standing as communal witnessing replaces private inspiration and insight as the final judge of truth claims. Certainly by the end of the nineteenth century, science and secularism were closely associated in their combined attacks against folk psychology, superstition, and religious revelation, each of which was replaced with a different way of reasoning (Chadwick 1975; Wilson 1999). Scientific knowledge thus sought to displace *opinion* in every realm of inquiry, including, of course, the human sciences. It is here we appreciate positivism's most general appeal.

The Human Sciences

Positivism as an ideal philosophy of science also found application in the human sciences. Note how John Merz in his influential (and magisterial) review of nineteenth-century thought summarized the situation at the end of the century:

> Clearly, besides the abstract sciences, which profess to introduce us to the general relations or laws which govern everything that is or can be real, there must be those sciences which study the actually existing forms as distinguished from the possible ones, and "here" and "there", the "where" and "how", of things and processes, which look upon real things not as examples of the general and universal, but as alone possessed of that mysterious something which distinguishes the real and actual from the possible and artificial. These sciences are the truly descriptive sciences, in opposition to the abstract ones. They are indeed older than the abstract sciences, and they have, in the course of the period under review in this work, made quite as much progress as the purely abstract sciences. In a manner, though perhaps hardly as powerful in their influence on practical pursuits, they are more popular: they occupy a larger number of students; and inasmuch as they also comprise the study of man himself, they have a very profound influence on our latest opinions, interests, and beliefs—i.e., on our inner life. (1896/1965, vol. 2, 203–4)

These human sciences, or what we now call the social sciences, followed the natural sciences, albeit in a more self-conscious fashion (Smith 1997, 2007).

The aspiration for scientific objectivity became a central concern for the social sciences when Auguste Comte proclaimed his positivist ideals in the 1820s (Comte 1825/1974). The enthusiasts argued that scientific methods were applicable to all domains of human need and asserted scientism as representing the best way to construct a worldview (albeit utopian) from one end of human experience to another—defining reality and comprehensively characterizing human psychology and sociology. (That ideology continues to have influence. For an example, see Wilson 1998.) Ironically, scientism aped its erstwhile opponent, religion, itself becoming a new religion, a new way of understanding *everything*, and in its unbridled enthusiasm, it threatened the entire scientific endeavor.

In the German context, "neutrality" served as the means of protecting the autonomy of the new science of society. And neutrality meant that "human values"—encompassing the entire spectrum of political, personal, religious beliefs and commitments—would not enter the analysis. Accordingly, science required dispassionate, neutral observation and interpretation. Anything less would putatively bias findings, interpretations, theories, and so on. Indeed, neutrality and objectivity went hand in hand into the laboratory. Investigations deemed nonneutral relative to some personal or social agenda were disqualified as "scientific." Thus "value-free" science and "neutrality" were regarded as, if not synonymous, at least closely overlapping. With this underlying motive defining the science/nonscience boundary, the emergence of the social sciences, more specifically sociology, had ready-at-hand criteria for its professional standards. As the founders of German sociology assembled themselves into professional groups, the "controversy over values" became a central issue. While virtually all agreed that identification with the natural sciences seemed crucial for sociology's legitimization, the boundaries of the discipline remained contested. The appeal of political involvement was enticing: the social sciences might be employed not only to rationalize social policy and promote the economy, but also to maintain social order and legitimate political decisions. But while some saw sociology as the answer to Germany's social and economic woes, others stiffly resisted. Among those of the latter group were the key architects of the new discipline, most notably Max Weber, Germany's foremost sociologist.

Weber clearly demarcated his role as scientist citizen, and thereby sought to protect the neutral (scientific) standing of the discipline by self-consciously distancing it from political movements, specifically socialism and Marxism. This was a contentious matter, since proponents of various political persuasions attempted to enlist the new social sciences to promote their own social ideals. But Weber was wary of applying biology to research programs devoted to proving theories of racial hygiene and social Darwinism. He argued that the so-called data was but "unbounded subjective valuations" (Proctor 1991, 111).

In "'Objectivity' in social science and social policy," Weber drew critical distinctions between the natural and social sciences by emphasizing the value-laden criteria of cultural and historical evaluations. Social science "involves subjective presuppositions insofar as it concerns itself with those components of reality which have some relationship, however indirect, to events to what we attach cultural *significance*" (Weber 1904/1949, 82; emphasis in original), and "on this significance alone rests its scientific interest" (81). "We cannot discover what is meaningful to us by means of 'presuppositionless' investigation of empirical data. Rather, perception of its meaning to us is the presupposition of its becoming an *object* of investigation" (76; emphasis in original). Any semblance of order is achieved by the winnowing power of value-based selection, for "in every case only a *part* of concrete reality is interesting and *significant* to us, because only it is related to the *cultural values* with which we approach reality" (78; emphasis in original). Such investigations made under the guidance of values, which direct selection and ordering of phenomena, are "entirely different from the analysis of reality in terms of laws and general concepts" (77), and more, laws are "meaningless" for the work of social scientists.

Weber advocated self-reflection and self-understanding of how incipient values and individual perspective influence interpretation of empirical evidence. Only "will and conscience, not empirical knowledge" can hold the scientist to the "ultimate standards" of objectivity (54). As Robert Proctor observes (1991, 135), Weber raised the banner of neutrality against the positivism of Comte and the developmental theses of Marxism (or for that matter, any of the "ideal types" of social theory [Weber 1904/1949, 106–10]). Weber argued that each erred in assuming a progressive, unique, and unilinear progression, the former in terms of modes of thought, the latter in terms of economic transformations. Both "wrongly assumed that there is some necessary relation between 'what is' and 'what ought to be.' In fact, Weber argued, the moral order cannot be derived from the movement of history or any other facts of social life. The real is not the rational . . ." (Proctor 1991, 135).[13]

From this neutral platform, Weber offered a general philosophy of the value question, one that remains cogent today. Protecting sociology's insularity from various political agendas, he responded with an articulate defense of value-neutrality for the new discipline. A scientist might involve himself in public debate by (1) identifying the contradictions between a person's values and their interests, (2) asking what empirical means are required to achieve those ends, and (3) pointing out the unintended consequences of pursuing those ends:

> These and only these can the scientist, qua scientist, undertake. The first is a question of logic; the second and third questions of empirical fact. But whether these goals should be pursued in the first place, Weber says, is a question no science can answer. For "this is a question of conscience, of personal commitment, and in this realm, science cannot tread." (Proctor 1991, 88–89)

Later, in his influential essay "Science as a vocation" (1919/1946), Weber shrewdly extended his espousal of neutrality for science in a bidirectional manner—science would not be contaminated with ideological bias and science, in turn, would not offer its own "objectivity" to matters of interpretation. Weber, keenly cognizant of the precarious status of the social sciences as science, attempted to delimit the erosion of an objective ideal, which he readily admitted was unattainable.

Today, the positivist agenda for the social sciences is no longer generally accepted (see Proctor 1991, 163–81, for discussion), and revised notions of how the social sciences might be based on discovery of their own general laws persist as a central theoretical concern. One strategy is to view such an inquiry as falling on a continuum of scientific explanation, where a more relaxed view of laws from that espoused in the natural sciences might suffice (e.g., Kincaid 1990). This line of argument often leads to a suggestion that a new *kind* of science is required, which then serves as a loose rationale for many programs that do not even attempt to cloak their efforts in the garments of scientific legitimacy. This admission weakens the entire edifice, prompting some critics like Lee McIntyre to maintain a stauncher stance (1993, 2006). However, for our purposes, the issue is not so much the conceptual basis of the social sciences and the quest for some firm scientific principles which might guide social policy, but rather the ever-present challenge of even understanding that agenda. In the following chapters we will examine several aspects of this general problem, directing our attention, first to the demise of the positivist ideal, and then tracking the development of a revisionist philosophy—one decidely less ambitious and more circumspect of its goals.

3

The Fall of Positivism

For my part I do, qua lay physicist, believe in physical objects and not in Homer's gods. . . . But in point of epistemological footing the physical objects and the gods differ only in degree and not in kind.

W. V. O. Quine, "Two dogmas of empiricism"

During the last century, global characterizations of science fell into three general groupings. The first cluster concerned itself with the placement of science within a general philosophical context, which meant interpreting the methods, products, and intellectual structure of science as part of a comprehensive epistemology. Most prominently articulated by Edmund Husserl, Whitehead, Heidegger, and John Dewey, critical themes emerged about how science framed the modern world in every aspect of human experience and how that presentation distorted or imperialistically trumped other forms of knowing. For convenience (and for reasons that will become apparent in the conclusion), I will refer to these diverse characterizations as "agent-centered." The second set, largely dominated by the logical positivists of the Vienna Circle but including earlier works of Pierre Duhem and Henri Poincaré, regarded science analytically by attempting to formalize the nature of observation, theory construction, and the basis of truth claims. They promoted the scientific enterprise by enthusiastically building on the foundations established by Comte, Whewell, and John Stuart Mill. Of the two approaches, at least in the

Anglo American community, the success of the positivist movement was self-evident by mid-twentieth century when Max Horkheimer (a founder of the Frankfurt School, and thus a Marxist in his orientation) opined—fairly, I think:

> Today there is almost general agreement that society has lost nothing by the decline of philosophical thinking, for a much more powerful instrument of knowledge has taken its place, namely, modern scientific thought. It is often said that all the problems that philosophy has tried to solve are either meaningless or can be solved by modern experimental methods. In fact, one of the dominant trends in modern philosophy is to hand over to science the work left undone by traditional speculation. Such a trend toward the hypostatization of science characterizes all the schools today called positivist. (1947/2004, 40)

Those who resisted positivism's advance or doubted its promises were characterized as suffering from "a failure of nerve" (Horkheimer 1947/2004, 40), and when Husserl lamented that "positivism, in a manner of speaking, decapitates philosophy" (1935/1970, 9), he obviously was referring to a different species of philosophy.

As different as these divergent approaches might be, the two contenders held in common a basic precept, for each assembly of critiques implicitly accepted that science claimed for itself a special form of reason and knowledge acquisition. A different, third major course in characterizing science appeared in the last four decades of the twentieth century, one that radically disputed this very claim and thereby rejected the unique status of science. Led by Kuhn, Feyerabend, and like-minded sociologically oriented critics, the adherents of this group believed that the practice of science and the production of knowledge followed a pragmatic course directed by unacknowledged social, political, economic, and cultural influences. Accordingly, they declared the positivists' formalization project moribund. These later thinkers summarily rejected those aspirations and their various accompaniments as suffering irredeemable flaws, not ostensibly as part of a neoromantic awakening, but rather as part of another agenda altogether. For simplicity, let us call it "postpositivist." Taking three nodal positions—agent-centered, positivist, and postpositivist—for orientation, this chapter examines the basis for the last of these positions.

To announce that we live in a postpositivist era hardly draws attention, but what such a declaration portends remains a beguiling issue. At least if science in its positivist mode has been "democratized," then a new political order has assumed primacy. Having dictated for over a century the criteria of true knowledge, the promise of material utopia, and the justification for belief in a rational secularism, science now must offer its knowledge as fallible, its material product as potentially dangerous, and its secular idealism as inadequate to address the resurgent fundamentalisms of our age. A new understanding replaced positivism, and with the ascension of the postpositivist contender, we must both acknowledge that change and understand how it alters virtually everything else.

This chapter surveys the rise of a radically different view of science. It does so with an eye on the status of the fact/value distinction as a guiding framework for understanding the shifting meanings of objectivity and the related problem of constructivism. The discussion probes a specific question that has appeared again and again in different formats as philosophers, historians, and sociologists of science grapple with the problem of how to depict science from their respective points of view: how do facts and values relate to each other in the context of characterizing the epistemology *and* social activity we call "science"? The spectrum of this matter spans the character of objectivity to the subjectivity of the knowing agent and invokes debates about the social construction of knowledge, the strictures of language, and the boundaries of science in its political, social, and economic environments. These issues lie at the foundation of science's truth claims and the process by which claims are made.

According to the revisionist approach, science was not entitled to any special claims based on some rigorous rationality. Its linguistic structure and strictures were as restrictive as any other cognitive activity. Most damaging to the underlying idealism of a logical rationality was the resulting relativism that seemed to undermine science's authority. In short, the demarcation problem shifted from the problem of distinguishing science from nonscience to one that considered how the new sociology of scientific knowledge diffused, or even eliminated, such differentiations. While the history of this transition is highly convoluted and combines many tributaries of thought, this project gained momentum from two seminal works published essentially at the same time: Kuhn's *Structure of Scientific Investigations* (1962) and Polanyi's *Personal Knowledge* (first appearing in 1958 and revised in 1962). We begin with the latter.

Polanyi: Personalizing Knowledge

Whether posed as the conflict of science and religion, as the competing claims of disjointed subjective and objective realms of experience, as the character of human nature and the status of the emotions, or as the moral standing of nature, a humanistic interpretation of an objective worldview remains a pervasive challenge for modernity. By tying together the various strands of this abiding tension, Polanyi offered a cogent presentation of this dilemma just as the positivist crest was about to crash. *Personal Knowledge* begins with the bald assertion, "I start by rejecting the ideal of scientific detachment" (Polanyi 1962, vii), and proceeds by analyzing the word "knowing" to show that its connotations refer to many levels of understanding. Impersonal, "objective" knowledge is only one kind aspired to, but even this category, according to Polanyi, is a conceit, and a limiting one at that. His argument attacked the positivists' position essentially from within the strictures of their own logic (which was, incidentally, very different from the strategy that Kuhn employed). I will only highlight certain aspects.

Much of Polanyi's critique concerned the logical futility of establishing any fixed framework that could critically test the positivist program. In other words, the positivists offered no perspective from which their own axioms might be examined critically. Specifically, we cannot escape our own perspective, the personal assessment that is intrinsic to any knowing. Simply put, Polanyi regarded the positivist view of science's logic as too narrow. He, like Husserl and Nietzsche, saw "rationality" as a broader category than the criterion of objectivity construed in a narrow sense. He observed:

> the act of knowing includes an appraisal; and this personal coefficient, which shapes all factual knowledge, bridges in doing so the disjunction between subjectivity and objectivity. It implies the claim that man can transcend his own subjectivity by striving passionately to fulfill his personal obligations to universal standards. (Polanyi 1962, 17)

Polanyi explicitly discounted *subjectivism* and substituted *personal.* In this fashion, he still aspired to objectivity's ostensible goals. This was not an either/or choice, for Polanyi would simply broaden the cognitive category of "objectivity" to include those mental faculties which play in the realm of discovery and cannot, in any formal fashion be finalized in logical format. He also explicitly recognized the "legitimacy of pre-theoretical experience—which is not the same as random subjectivity!" (Hansen 1990, 14). He called this broadened realm of knowing the "tacit dimension" (Polanyi 1966), and in that domain the full panoply of knowing—aesthetic sensibility, probabilistic judgment, intuition, metaphoric extension, and the like—comes into play. Briefly, Polanyi argued that we see the world through different cognitive lenses, each of which has a part to play in scientific discovery. In still offering an objective vision of the world mediated by the active person in his or her various knowing modalities, Polanyi resurrected the deeper metaphysical goals of science. He would employ objectivity as a humane tool:

> Objectivity . . . does not require that we see ourselves as a mere grain of sand in a million Saharas. It inspires us, on the contrary, with the hope of overcoming the appalling disabilities of our bodily existence, even to the point of conceiving a rational idea of the universe which can authoritatively speak for itself. It is not a counsel of self-effacement, but the very reverse—a call to the Pygmalion in the mind of man. (Polanyi 1962, 5)

For Polanyi, science is a passion, which, despite its apparent austerity and aloofness, must reflect a deeply personal way of viewing the world:

> [P]ersonal knowledge in science is not made but discovered, and as such it claims to establish contact with reality beyond the clues on which it relies. It commits us, passionately and far beyond our comprehension, to a vision of reality. Of this responsibility we cannot divest ourselves by setting up objective criteria of verifiability—or falsifiability, or testability, or what you will. For we live in it as in the garment of our own skin. Like love, to which it is akin, this commitment is a "shirt of flame," blazing with passion and, also like love, consumed by devotion to a universal demand. Such is the true sense of objectivity in science . . . the discovery of rationality in nature, a name which was meant to say that the kind of order which the discoverer claims to see in nature goes far beyond his understanding. (1962, 64)

Echoes of pre-Socratic *logos*, the Reason that underlies reality and governs all physical and organic nature, may be heard in Polanyi's almost religious testament. Wary of becoming ensnared in the confines of restricted theory or disciplines of thought and, more importantly perhaps, limiting scientific method to only a narrow wedge of experience and modes of knowing, Polanyi thought that the scientist should again visit the dilemma of what warrants inclusion in the scientific domain. The problem of integrating different layers of reality coupled to the endeavor of widening the scope of investigation would then become a challenge of devising inclusive cognitive criteria that would loosen the strictures encasing notions of science held by his contemporaries.

Polanyi did not revive subjectivism, but rather espoused subjectivity's recognized role in scientific discovery and theory formation. Instead of denying the selective process of observation and the interpretative character of scientific investigation, Polanyi embraced them. Thus "personal knowledge," became a catchall for the necessary, creative elements that cannot be accounted for in the positivist rendition of science. For Polanyi, objectivity is intrinsically coupled to notions of subjectivity: one cannot speak of objectivity without at least implicit reference to its counterpart. Said differently, the subjective has been recast. This perspective is hardly novel, nor restricted to an eccentric. For example, Lewis Wolpert and Alison Richards (1997) have amply documented (again) among a diverse group of practicing scientists how subjectivity is at play in any assessment—whether wonder, curiosity, passion, pleasure, disappointment, competitiveness, aesthetic appreciation, and so on. This spectrum of emotions appears in the scientist's everyday work—from postulating new experiments to their final public judgment—and, despite the positivists' valiant attempt to formalize scientific discovery or verification, these "contaminating" subjective elements require an accounting.

Polanyi's project expands postpositivism into realms vacated by objectifying positivism. Accordingly, scientific findings alone are insufficient for determining significance, and thus interpretation is required. Indeed, this insight has a long history. Raw knowledge, a fact, is essentially meaningless. What is the significance of a scientific fact or larger theory unless we may apply it to human understanding? "Understanding" entails many layers of interpretation, and here we see the linkage to recent constructivist arguments. Science influences its supporting culture, the values that govern its use, and, ultimately, the sense of meaning and significance ascribed to the scientific portrait of the world. Polanyi thus recognized, as did an entire generation following him, that scientific knowledge was ultimately human centered in at least two senses.

First, Polanyi focused on the scientist as a unique knower. He specifically emphasized the judgmental and interpretative aspects of scientific thinking, which could not be adequately accounted for by some prescribed logic of scientific discovery. The creativity of the scientific imagination rested on many faculties, some "tacit" and thus buried, some implicit and thus undeclared, and still others considered subversive to science as normally understood (e.g., aesthetic judgment) and thus even more subtly operative. But he chastened those who denied their cognitive contributions. In other words, the simple inductive model—data in, conclusions out—could not capture the scientific process at the level of the individual scientist creating and interpreting her research.

Polanyi sought to bring the personal into the domain of science, or we might say, he endeavored to broaden scientific thinking to include the personal, without the radical de-legitimization of science itself. This was Weber's strategy (albeit undeveloped), and it has had a rich history, both in assessing the intersection of art and science and in attempting to decipher the cognitive or, more broadly, the psychology of scientific creativity. To be sure, a new movement in cognitive psychology has given strong impetus to this intuition by describing multiple forms of "emotional intelligence." Accordingly, beyond IQ, it appears that the successful ability to navigate the world also depends on such measurable qualities as social sensitivities and self-awareness; self-control and self-projection; motivation and empathy. Such acumen contributes to the ability of effectively applying analytical aptitude and the other multiple dimensions of cognition (Nussbaum 2001).

By emphasizing what had heretofore been referred to as emotional characteristics, Polanyi presciently identified and promoted faculties of knowing that have become key issues in contemporary cognitive psychology.[1] In the scientific setting, intelligence is typically regarded as a paramount cognitive virtue, yet we might well ask, what kind of *intelligence*? Undoubtedly, analytic ability and its various cohorts are crucial for success, but effectiveness also requires judgments that are emotionally based, or at least inspired.[2] Emotions color evaluations based on the context of their expression, the web of beliefs in which they are situated, and these elements, as Polanyi himself observed, typically remain *tacit* (Polanyi 1966). Indeed, Quine noted that justification for theory choice or determination of relevant information entails a selection rarely understood in any rigorous "rational" fashion (Quine 1953a/1961/1980; discussed below). This suggests that emotions can render information salient, enabling selection of some and thereby weaning the overwhelming influx of the rest (de Sousa 1987). The theme that evaluation of epistemic judgments reveals the participation of extrarational subjectivity extends forcefully from Quine's critique of the logic of positivism, to Kuhn's study of scientific revolutions, to Ronald de Sousa's *The Rationality of Emotion* (1987).

Because of the holistic character of decision and belief formation, we must also factor in the role of emotion as a cognitive faculty in filling gaps in reason and perception (Hookway 2002, 259). Such patterns of salience that emotional processing provides can then function "as 'a source of reasons,' as something that comes '*before* reasons'" (260). Thus, while some emphasize the aesthetic influences on decision making (Tauber 1996b), de Sousa perceives that emotion more generally establishes patterns of relevance among objects of attention and inferential strategies. Others pose the balance more abstractly as a complex mixture of "informed intuition based on emotional coherence" (Thagard 2000). However relevance is established or balance attained, all commentators are seeking the means by which the multiple elements of cognition are assimilated and action prioritized. Herein lies the central point: coherence of thought requires the integration of *all* those elements that go into thinking—traditionally conceived rationality *and* emotion (the latter of which is all too often understood as "irrational").

To separate scientific rationality from other components of intelligence as some distinct and independent ability distorts the process, for thinking in all of its various formats requires emotional judgment. Our web of beliefs (discussed below) is diverse (Nelson 1990), and any representation has attached to it an "emotional valance" (Bower 1981) that must be factored into the calculus of judgment. While coherence has been formalized (Thagard 2000), it seems plain that at this level of intuitive emotional content, quantitative designations are highly improbable. Nevertheless, appreciating that emotion plays its own role in the objectivity of scientific research suggests that the unified reason we seek already resides in the ongoing project of understanding how integration of various faculties of knowing might comprise a more comprehensive theory of reason. Perhaps more to the point, that quest seems justified on the merits of understanding the scientific process in its full employment.

This perspective has a venerable history. Before Polanyi, Kant designated "judgment," Goethe perceived the archetype (Tauber 1993) and Nietzsche championed "art" in their respective attempts to characterize the complex interplay of analytical and emotional faculties that result in achieving insight. Each argued for the aesthetic faculty as accounting for a critical component of human intelligence (Tauber 1996b), while contemporary commentators have emphasized broader categories of consideration. In any case, we may draw a direct line from Goethe to Polanyi and beyond: human imagination exercised in its various modalities is linked by diverse forms of human creativity. After all, the line separating objectivity from subjectivity is highly dynamic, historically contextualized, and continuously contested. The romantics placed this problem front and center, whether addressed in philosophy, poetry, or psychology. Certainly, how to integrate different kinds of intelligence reflected the deeper metaphysical issue of how to make the world whole—whole in the sense of placing the individual *in* the world, as opposed to peering *at* it as a spectator. Polanyi readily falls into this tradition because for him, the scientific endeavor offers a worldview that not only supplements other ways of knowing, but does so as a *result* of drawing upon those sources.

Personal meaning is the union of objective knowledge and subjective modalities of various kinds. The scientific project has expanded the reaches of time and space, the complexity of the natural world, the order of natural processes, and the evolution of nature. And each domain of inquiry has offered a picture of the world in which a systematic study provides not only factual knowledge but also a world that must be interpreted and integrated into a personal worldview, one filled with meaning and signification. Interpretation originates in the same metaphysical wonder that stimulates scientific inquiry in the first place. On this view, not only is some portion of the scientific enterprise beholden to subjective elements, an emotional motivation rests at the base of the entire enterprise. Therefore, beyond the aesthetic components at play in scientific judgment, the analytic and logical skills of the scientist ultimately serve some existential function as well. Although such psychological factors are rarely discussed, they remain insistently present, albeit largely unaccounted for in our comprehension of scientific reasoning (e.g., Maslow 1966; Faust 1984; Root-Bernstein 1989).

Kuhn: Raising the Lid of Pandora's Box

Prior to 1960, what passed for philosophy and history of science we now recall as the *Standard View*. In large measure a hagiography of "scientific method," "scientific rationality," "scientific objectivity," and "scientific progress," the Standard View portrayed science as logical in its ordered definition of the real. Despite the positivists' best efforts, when each of these categories was placed under a critical lens, it was found to fail their own cognitive standards. Indeed, "the idolatry" (Zammito

2004, 52) of science as an extraordinary mode of knowing, protected by self-correcting mechanisms from fallacy and bias, shattered upon critical appraisals. The most celebrated of these works was Kuhn's *Structure of Scientific Revolutions* (1962, 1970), which more than any other work crippled the scientism of the era.

Ironically, *Structure* originally appeared as one of several long essays in the logical positivists' project *Foundations of the Unity of Science: Toward an International Encyclopedia of Unified Science*, whose nineteen monographs Otto Neurath, Charles Morris, and Rudolf Carnap edited jointly. That project was part of the thesis shared with earlier forms of positivism, that all sciences might be unified under universal principles.[3] The natural sciences, of course, served as models of such knowledge, and the irony of Kuhn's work is not only that *Structure* appeared under these auspices, but that it was interpreted sympathetically perhaps because he, in some sense, had universalized the sociology of scientific knowledge, albeit without anticipating the effect it would have in undermining the entire positivist program.[4]

If Polanyi built a cabin (à la Thoreau), Kuhn erected a mansion. By radically challenging the notion of science proceeding in some stepwise, rational fashion, Kuhn cast into doubt the very logic and objectivity of science the positivists averred. The view of the autonomous, rational growth of scientific thinking—that is, of science as logically progressing and possessing universal and unwavering objective criteria to describe nature—is perhaps a product of conflating science's declared ideal aspirations with the more subjective and heterogeneous nature of its enterprises. The idealized description begins with the unquestioned acceptance of science as advancing by its rational constructs and achieving its success through some unbridled objectivity. In refutation, Kuhn maintained that scientific change occurred nonincrementally in sudden leaps, or what *Structure* described as "paradigm" shifts. Kuhn's notion of paradigm—a term now hopelessly weakened by overuse and lost amongst its various interpretations (Masterman 1970)—represented this perceptual-conceptual holism of fact and theory, the gradual or incremental changes of which cannot account for major scientific revolutions. From this perspective, final truth forever retreats, leaving only facts that cohere in the theoretical, technological, and methodological arrangements of the time. In other words, as already discussed, facts cannot claim any autonomous standing independent from particular paradigms in which they are placed. Facts only become facts within their encompassing theory, much as Goethe had observed at the end of the eighteenth century.

The picture Kuhn portrayed led to an intense historical exploration of specific instances of scientific change, as well as the more general problem of characterizing scientific development. Prior to *Structure*, the historiography promoted a simple normative view. The so-called *Growth Model* of the eighteenth century persisted into the twentieth century in the work of George Sarton, who maintained that science advanced by rationally exact methods, reflecting the unity and continuity of knowledge. Independent of social pressures and historical contingencies, science was regarded as progressing as it discovered the Real World (Richards 1987). The *Revolutionary Model,* first advocated at the end of the eighteenth century, emphasized the sudden shift of thinking inaugurated by the Scientific Revolution inaugurated by Bacon, Galileo, and Copernicus (Butterfield 1957). *What* that revolution was and continues to be is open to interpretation (e.g., Shapin 1996; Cohen 1994), but *that* science did indeed undergo a radical methodological and metaphysical shift in the sixteenth century is generally acknowledged (e.g., Koyré 1957). Whether growth or revolution, each of these views regarded science as essentially progressing within its own system of rational inquiry, fulfilling its mandate to describe nature objectively.

Most scholars would agree that history and philosophy of science began as part of science proper, and thus the histories were governed by the self-image scientists themselves offered of their enterprise. Indeed, prior to our own century, history of science was primarily a rhetorical and theoretical tool in showing how new science was part of a progressive and rational process. Review of the historical development of a particular science was an integral component of the scientific report. When Goethe wrote on color theory, Joseph Priestly on electricity, and Charles Lyell on geology, they used history to legitimate their own work. Even into our own era, history of science (when still entertained as relevant to science) was often seen as exercising a beneficial influence on practice. For example, the laboratory scientist might profit from history used as an analytical tool (Kragh 1987, 33–34). Although the historical perspective might be valued and certainly governed such innovators as Giambattista Vico, confusion about historical interpretation as an important scholarly activity distinct from laboratory science and ongoing theory construction was untangled slowly.

Kuhn changed the frame of reference. *Structure* based its radical inferences from a simple observation: scientific memory is short. The depreciation of historical fact is deeply, and probably functionally, ingrained in the ideology of the scientific profession, the same profession that places the highest of all values on factual details of other sorts (Kuhn 1970, 138). This important insight generated at least two ideas: The first was the fundamental question of how and why science proceeds without self-consciousness of its own method. The second—and the one Kuhn himself, and later the entire discipline, pursued—concerned the extent to which science might be characterized by some historical self-reflexiveness. The first set of questions would in effect be answered as a shadow response to the second set.

According to Kuhn, the practicing modern scientist had little, if any, historical consciousness, because the more historical detail, whether of science's present or its past, or taking responsibility to attend to the historical details presented meant highlighting human idiosyncrasy, error, and confusion. Why dignify what science's best and most persistent efforts have made it possible to discard? The answer proved an embarrassment. However, upon critical examination, the history of science exhibited no logical, progressive rule or method. Instead it evolved by, what appears from later perspectives as, convulsive, sudden revolutions that can hardly be accounted for by a simple narrative that describes incremental, stepwise, conceptual changes governed by some independent, sacrosanct rationality.

In parallel, philosophy of science was irrelevant to the practicing scientist (Kuhn 1970, 86). Accordingly, only in crisis periods of revolution would scientists seek out the rules and assumptions of their work. In "normal science," the research edifice stands stable, and scientific self-consciousness is not required and thus not sought. (*Structure* differentiates "normal" science, where there is progress in an accumulative fashion within a given paradigm from those sudden conceptual changes that generate a "revolution.") In fact, the relative distance, or objectivity, required for such assessment simply has no function in the everyday consciousness of the working scientist. In short, to do science, and so to be a successful scientist, historical or philosophical awareness is unnecessary. And just as a lawyer need not know sociological theory of small groups to practice law before a jury, so a scientist hardly requires a sociological analysis of his own competitive group to function successfully within it.

The notion that scientists take little heed of the historical context of their research became an adage, even a precept, of science studies. To the extent that scientists address the past, they constantly refashion the antecedents of their work in the image of contemporary prevailing theory and explanation. Thus an objective historical account of the development of scientific understanding or an acknowledgment of possible nonrational factors at play in the evolution of the scientific discipline accordingly plays no significant role in the professional scientific endeavor. In other words, self-reflection on scientific practice has become a specialty of others, while scientists for the most part restrict themselves to *doing* science as best they might.

But assuming a perspective from afar, a novel question presents itself, or at least so Kuhn argued. He asked, if science, to proceed, does not rely on understanding its own origins and tracing its historical evolution, then to what extent does it adhere to such rational categories of development? The answer he offered—and later aggressively pursued to the point of his own repudiation—argued that scientific *evolution* was a process governed by many factors, of which the objective characterization of nature hardly stood alone. Famously, paradigms were the conglomerate of cultural, political, economic, aesthetic, and sociological ingredients that became additional contributing factors to what had hitherto been solely scientific rationality. Further, and even more disturbing, the pursuit of truth remained only an ideal, for *truth*, *rationality*, and *objectivity* evinced by the historical record were contingently constructed. Albeit subject to the restraints of nature and highly flexible in its interpretative schemata, science lost its assumed logical footing. Uncertain objectivity yielded to constructivism, bequeathing the problem of defining interpretative schema.

Kuhn's *Structure* itself became a paradigm of sorts by replacing normative points of view with descriptions emphasizing the social interdependence of scientific practice and theory, the philosophical uncertainty of scientific methods, and the spasmodic record of scientific change. Kuhn's interpretation is a *Gestalt Model* (Richards 1987), that is, scientific paradigms shift so that the competing worldviews appear as distinct pictures. To illustrate, take, for example, the famous duck-rabbit drawing where a single image may appear as either a duck or a rabbit. At first glance, most people see either one image or the other; that is, one animal appears to them as distinct and independent, and, if prompted, they are able to see the other animal. One image or the other holds the viewer's attention, namely, the duck and rabbit cannot be seen together. Similarly, Kuhn argued that when we look at the world through paradigms, we behold a particular picture, namely, that we become "locked into" a paradigm and thereby fail to "see" (or shift) between that view and an alternative. Indeed, divergent scientific theories, while referring to the "same" phenomenon, assemble those phenomena in radically divergent ways. According to this gestalt thesis, the differences of Newtonian and Aristotelian science disallow translation of their respective descriptions from one system to the other. In other words, they are *incommensurable*, not just different.

The gestalt picture of scientific change models how context, past experience, and familiar assumptions control our perceptions and conceptual experiences. On this view, the metaphysics in which we are enveloped preclude living within another system or even translating that system into terms that are discernable to our own fundamental understanding of *what* the world is and how it basically works. The cardinal principle harkens back to Goethe's original observation that facts are theory-laden. Simply, *what* scientists see depends on the theories or models with which they work.[5]

Kuhn's key concept, and the one generating the most discussion, was not his argument concerning scientific change (i.e., a skeptical assessment of the picture of an orderly progression of scientific growth), but that such revolutions may render the vanquished paradigm incommensurable with its successor. Here the radical character of the presentation holds full force: if one worldview is incommensurate with another, humans are essentially limited to the metaphysical world that they inherit. So powerful is such a view that the previously accepted theory essentially captures its followers within its epistemological and ontological web. Under such circumstances, how might objectivity be attained? Indeed, what is reality if the mode of description competes against other contenders with radically different precepts and presuppositions (those that cannot be challenged) that hold its believers firmly within their grasp? The merits of the case have ranged from the technical limitations imposed by instruments of inquiry to the arbitrary conceptual boundaries science must accept in its pursuit of "truth." (See Hoyningen-Huene 1993 for exhaustive references regarding Kuhn's argument.)

Kuhn's own views were modified to the point of disavowing the "Kuhnians." In two later essays (Kuhn 1991, 1992), he reformulated his position regarding scientific knowledge, and more specifically incommensurability, in an endeavor to salvage *Structure* from a radical relativism. Kuhn built his case from familiar material: science firmly embedded in its culture ("The Archimedean platform outside of history, outside of time and space, is gone beyond recall"), an evolutionary epistemology ("Scientific development is like Darwinian evolution, a process driven from behind rather than pulled toward some fixed goal to which it grows ever closer"), and an antirealist orientation for truth claims ("If the notion of truth has a role to play in scientific development, which I shall argue elsewhere that it does, then truth cannot be anything quite like correspondence to reality" [1992, 14]). From these positions, Kuhn redefined incommensurability as a conceptual disparity between specialties that have grown apart, a "sort of un-translatability, localized to one or another area in which two lexical taxonomies differ" (1991). In some sense, he continued to share the same positions advocated by his more radical followers, and he could never disassociate himself from the radical realignments that he spawned.

The historians following Kuhn developed new historiographic models, which included more interest-driven scenarios, for example, social-psychological models describing science as fueled by social interests and psychological needs to the point of driving, if not determining, theory conceptualization. Marxist, Darwinian, feminist, and various constructivist orientations followed (Richards 1987). Finally, those attempting to find a more comprehensive scaffolding turned to a variety of evolutionary models, represented most prominently by Gerald Holton, David Hull, Imre Lakatos, Karl Popper, Robert Richards, and Stephen Toulmin. On the evolutionary view, theory survives not solely by appeal to evidence but because other competitors are less fit. The intellectual environment selects ideas and restricts those that might otherwise be entertained in a complex calculus of the intellectual traditions and the social situations of its practitioners. Science then is viewed as an ever-evolving enterprise, the business of which is to attend to the picturing of the world. However, as representation it only achieves an approximation. The pragmatic epistemological judgment of its practitioners is always unsteady as they probe to verify their depictions with nature as best they can. This theme recurs throughout contemporary science studies, so, besides stimulating competing models of scientific change, *Structure* spawned an entire generation of science studies that is inspired by its central thesis. That the book was later co-opted by strong constructivist Kuhnians,[6] and other postmodernists, whom he shunned, is only one of several ironies, which include his professional homelessness, as he sought acceptance by philosophers of science who chose to ignore him.[7]

Despite the undoubted influence of its argument, the ultimate influence of *Structure* more likely rests on deeper philosophical arguments and other shifting elements in the study of science. By 1960, the positivist program was already unwinding from within (Friedman 1999), and Kuhn, for all of his originality, drew upon a profound philosophical reassessment of positivism led by Quine. How Quine's insights have been extended to the foreground of science studies cannot be overemphasized, for he, more than any other figure, unhinged descriptions of scientific practice and its pragmatic logic from formal accounts.

Quine and the Dismantling of Logical Positivism

For twentieth-century positivists, that is, the logical positivists, language became the arena in which to examine science's philosophy. They maintained that scientific method is the only source of knowledge, and that a *statement* is meaningful only if it is "scientific," in other words, empirically verifiable. (For our purposes, we will ignore the differences between "logical" and "empirical" positivists.) While discarding metaphysics as meaningless and thus something that could be ignored, they endeavored to ground science in language, which they had hoped would follow logical analysis (Ayer 1952).

The logical positivist movement (also called the Vienna Circle) of the 1920s arose, in large measure, as a revolt against the idealism that dominated philosophy at the turn of the century (see Reichenbach 1951; Frank 1949; and various papers in Ayer 1959). Joining other analytic philosophers, they rejected metaphysics tout court by specifically arguing against traditional metaphysics through the logical analysis of language (Hylton 1990; Giere and Richardson 1996; Tait 1997). Statements alluding to some transcendental reality were regarded as meaningless, since they could not be verified. Thus the knowledge criteria of science were extended to the domains that could not be so characterized, not only as beyond empirical science, but beyond analytic discourse altogether.[8]

In their analysis of language, the logical positivists pursued both a "negative" program and a "positive" one. The first dispenses with "nonscience" (a major focus of concern) by establishing a linguistic conception of analytic truth that would provide an account of the nonempirical character of logico-mathematical knowledge. Without appeal to metaphysical principles or abstract entities (like concepts or ideas), these positivists attempted to establish the a priori status of logic and mathematics compatible with a radical empiricism by showing the truth of such propositions through logical analysis. Having putatively secured logic and mathematics and having pushed metaphysics aside, positivist philosophy was then freed to do epistemology in the same analytical *linguistic* manner. Their philosophy thus became the analysis and clarification of meaning with the use of logic and scientific method. Accordingly, language was viewed as a system for solving problems; from another vantage, philosophical problems were characterized as confusions bestowed by language itself, or as Wittgenstein famously noted, "philosophical problems arise when language goes on holiday" (Wittgenstein 1953/1968, 19e). Accordingly, the aim of linguistic analysis is to solve philosophical problems, namely, "to shew the fly the way out of the fly-bottle" (103e). These efforts, however, failed (as Wittgenstein might have predicted given his suspicions [see n. 8]), and given this new opening, a spectrum of options ranging over varieties of naturalism, pragmatism, constructivism, and relativism have made their respective claims.

Logical positivism's failure had many sources, but for our present purposes the issue reduces to a single fault: for the logical positivists, the key to cognitive significance rested on mutually exclusive criteria, that is, logical and factual. Thus meaningful statements either were *analytic*—independent of empirical considerations and reliant on language alone (as Quine wrote, "grounded in meanings independently of matters of fact"—or *synthetic* (assertions which were verified or falsified by empirical procedures, in other words, "grounded in fact" [1953a/1961/1980, 20]). Indeed, the demarcations—theory/observation, discovery/verification, fact/value—rested on this more fundamental division between synthetic and analytic statements.

Mathematical and logical statements were regarded as analytical (tautologies) and true by definition. Such propositions are helpful in organizing cognitively meaningful statements, but are not verifiable by examining the world and thus say nothing about the world. In contrast, synthetic truths are empirical. The analytic/synthetic division so understood originates with Kant, who argued in the *Critique of Pure Reason*

(1787/1998) that sensory experience requires mental (cognitive) synthesis, while analytic statements are tautological and rest within their own internal logic and definition. For instance, the truth of the statement, "All unwedded men are bachelors," depends solely on the definition of "bachelor" and thus is an analytic statement. "I dropped the ball" is a synthetic statement. Its truth content is assessed by determining whether I actually dropped a physical sphere that bounces, and if not, whether my statement refers to having failed an assignment or responsibility or some other referent. In short, synthetic judgments require some interpretative, empirical operation and thus are distinguished from analytical statements. Or at least so it seemed.

The so-called analytic-synthetic distinction collapsed under Quine's critique, which showed that synthetic statements could not be completely separated form analytic elements that supported them. To say that "Caesar crossed the Rubicon" cannot suffice as a synthetic statement because the very meaning and significance of that sentence requires a vast network of supporting facts, definitions, and interpretations, which, in turn, create a web of beliefs. This "holism" set the stage for Kuhn's paradigm, because Quine argued that it is entire theories that hold empirical significance. This position had broad ramifications.

Quine effectively argued that theories are tested as ensembles, not singly. Because (1) any scientific statement can be held true if adequate revisions are made elsewhere in the system; and, conversely, (2) no statement is immune to change, truth claims are made within the context of the whole, and not even analytical statements are free of such adjustment (Quine 1953a/1961/1980, 1991; Laudan 1990b). As Quine wrote, "our statements about the external world face the tribunal of sense experience not individually but only as a corporate body" (Quine 1953a/1961/1980, 41). (Later, he would moderate the general holism, that is, "all of science," to the unit of empirical evidence.)

Two elements deserve emphasis. First, language (and by extension, belief systems and scientific theory) achieves stability by balancing all respective elements within a holistic construction (1953b/1961/1980, viii). The web of beliefs acts as a kind of buffering system for accommodating new elements and bestowing meaning on them by their coordination within the system as a whole (Quine and Ullian 1978). The second element describes the relativity of the process inasmuch as each system (language, theory) has its own coordinates, its own inner logic, its own posture relative to other systems. More formally, "Because reference is arbitrary, reference *is* nonsense except relative to a coordinate system" (Quine 1990, 7; emphasis in original). This simply means that language fits loosely to the world, and the manner in which words or statements link to the world is arbitrary and thus indeterminate, except as integrated within some "coordinate system." Quine dubbed this claim as the "inscrutability of reference" (Quine 1969a).

The Quinean limits of the analytic, and the ensemble of various indeterminancies (of reference, of underdetermination, and of translation [1990]),[9] placed science under a scrutiny that inaugurated a revolution in characterizing its truth claims and objective methods. Instead of some idealized notion of truth or the singular truth quotient of any particular fact, all of the elements of knowledge—facts, hypotheses, theories, the diverse values supporting each, the linguistic structures and metaphors, the larger social and cultural determinants, and so forth—contributed to what he called a web of beliefs. Like a web, any alteration of one part signified an adjustment that would either accommodate or reject that component. Once incorporated, all the other supporting elements would also have to adjust to the integration of the new part. In the last section of "Two dogmas of empiricism," Quine summarized his position better than any commentator:

> [T]otal science is like a field of force whose boundary conditions are experience. A conflict with experience at the periphery occasions readjustments in the interior of the field. Truth values have to be redistributed over some of our statements. Reevaluation of some statements entails reevaluation of others, because of their logical interconnections. . . . But the total field is so underdetermined by its boundary conditions, experience, that there is much latitude of choice as to what statements to reevaluate in the light of any single contrary experience. No particular experiences are linked with any particular statements in the interior of the field, except indirectly through considerations of equilibrium affect the field as a whole.
>
> If this view is right, it is misleading to speak of the empirical content of an individual statement—especially if it is a statement at all remote from the experiential periphery of the field. Furthermore it becomes folly to seek a boundary between synthetic statements, which hold contingently on experience, and analytic statements, which hold come what may. Any statement can be held true come what may, if we make drastic enough adjustments elsewhere in the system. (1953a/1961/1980, 42–43)

In repudiating the "imagined boundary between the analytic and the synthetic," Quine espoused a "more thorough pragmatism. Each man is given a scientific heritage plus a continuing barrage of sensory stimulation; and the considerations which guide him in warping his scientific heritage to fit his continuing sensory promptings are, where rational, pragmatic" (Quine 1953a/1961/1980, 46).

The pragmatic, local descriptive alternative Quine offered maintained that the reality sought by scientists was a metaphysical aspiration, discerned by substituting their linguistic analysis for a traditional metaphysics (Quine 1969b). He argued that we must be satisfied with the picture offered by our investigations but claim no more (Quine 1969c). Truth can only be defined within its particular framework. As Neurath famously noted, "We are like sailors who have to rebuild their ship at sea, without ever being able to dismantle it in dry-dock and reconstruct it from the best components" (1931c/1983, 92). Language then assumes the centrality of philosophical analysis:

> If we cannot get out of the "conceptual boat" represented by one linguistic framework or another and gain a clear view of the world as it really is—indeed, if it does not even make sense to imagine such a thing—then why not turn our attention toward the boat itself and the conceptions about reality it embodies? (Romanos 1983, 31)

Indeed, Quine followed Carnap and Neurath in recognizing that our conceptions of reality are determined by language, a language that cannot be escaped nor viewed from afar. No less than reality is at stake.

> The fundamental-seeming philosophical question, How much of our science is merely contributed by language and how much is a genuine reflection of reality? is perhaps a spurious question which itself arises wholly from a certain particular type of language. Certainly we are in a predicament if we try to answer the question; for to answer the question we must talk about the world as well as about language, and to talk about the world we must already impose upon the world some conceptual scheme peculiar to our own special language. (Quine 1953c/ 1961/1980, 78)

Neither language nor scientific conceptual schemes mirror nature, and thus assessing the success of any scheme is based on pragmatic criteria. These are adequate for the task at hand, albeit "truth" assumes a modest stance. The process is piecemeal, yet progressive. Picking up the boat repair metaphor again:

> Yet we must not leap to the fatalistic conclusion that we are stuck with the conceptual scheme that we grew up in. We can change it bit by bit, plank by plank, though meanwhile there is nothing to carry us along, but the evolving conceptual scheme itself. The philosopher's task was well compared by Neurath to that of a mariner who must rebuild his ship on the open sea. (Quine 1953c/1961/1980, 78–79)[10]

Accordingly, the belief system is not dependent on what is *really* there, but rather on the success with which it works. And it "works" through observation and the hypothetical-deductive method, which then offers a "conceptual scheme" of the real. And conceptual schemes, like frames of reference in relativity theory, serve as *our* perspective. *Reality* is then only *our* best theory. Metaphysics shifts to an ontological commitment to naturalism and its deserts. Truth then is a product of this pragmatic approach, an approach whose limits we better understand but whose success is beyond any final logical analysis.

Yet pragmatic results *do* arrive. According to Quine, science is an *object-oriented idiom,* and scientists "speak of objects" that ascend a ladder of public identification and abstraction. We are not in "cosmic exile," but approach the real with confidence, albeit conforming to a good measure of skepticism. In the end, and maybe inconsistently, Quine's naturalism balanced his relativism, and that naturalism was steadfast. Note that the position derives from a philosophical critique of language, just as the positivists insisted but hardly had foreseen:

(1) Language is not pictorially related to the external world and thus cannot provide an isomorphic depiction of the world ("correspondence" theories of truth are therefore forbidden). Thus language (propositions) cannot present us with a picture of reality as whole, nor can language represent (correspond to) some final or ultimate reality.

(2) Word meanings are derived from the context of their use (e.g., "I am going to the bank" may mean I am going to a river or I am going to get some money), and thus meanings must be considered in the universe of rules, context, habits, and conventions that bestow meaning.

(3) Different language functions must be differentiated (e.g., naming, classifying, commanding, prescribing, describing, referring, expressing, etc.) and not conflated.

And, perhaps most importantly:

(4) Language cannot go "behind" itself, since we would have to use either its own symbols or other symbols in an endless regress to describe language. Language thus offers us no Archimedean point in which to either describe language itself or the reality that language describes, thus we are left with the dilemma of understanding language's strictures as it serves as the vehicle of the mind's exploration of the world. Reality may be viewed in alternative ways, not because the nature of facts depends on how we construe or understand them to be, but rather because there are no such facts except relative to some linguistic or conceptual framework within which we live (Romanos 1983, 29). In short, according to the Quinean position (although much had been argued earlier by Neurath [see n. 10]), language cannot be beached. Our judgments are embedded within the boat of our language and concepts. Our language only changes piecemeal, slowly, with no new design available. The architectonics of theory are similarly trapped.

Understanding the basic contours of Quine's position and those he debated places the conceptual issues currently in dispute on their philosophical scaffold—namely, the constructivist position that defenders of a positivist science see as threatening reason, objectivity, and the pursuit of truth. Admittedly, this summary of Quine's views hardly does justice to the complexity of the issues he raised and reformulated (see Quine 1990 and 1991 for his own summary), but in appreciating how he served as a transition between the positivists and the later constructivists who followed them, we see a deep unresolved tension: on the one hand, his commitment to natural epistemology places him firmly within the realist camp, yet, on the other, his philosophy of language opens the way to antirealism.[11] And the net result? Well for most of Quine's followers, a pluralistic universe emerges. "What we must face is the fact that even the truest description comes nowhere near faithfully reproducing the way the world is. . . . I reject the idea that there is some test of realism or faithfulness in addition to the tests of pictorial goodness and descriptive truth. There are very many different equally true descriptions of the world, and their truth is the only standard of their faithfulness" (Goodman 1972, 29–31). What Goodman meant by truth has been an imbroglio of vast confusion,[12] but leaving that matter aside, suffice it to conclude that Quine's ontological relativism, affording only pragmatic criteria for knowing the world, represents a bedrock position to which later science studies constantly refer, either explicitly or, more often, implicitly.

Upon Quine's platform, Kuhn built an argument that pitted historians and sociologists against scientists who embraced a positivist view of the world. So, while Kuhn more directly influenced later science studies, Quine, more than any other philosopher, must be credited with disassembling the positivist program (e.g., Romanos 1983; Hookway 1988; Barrett and Gibson 1990; Nelson 1990; Friedman 1999; Zammito 2004). That he opened the door to radical reactions against positivism conjures a certain irony, since Quine's own naturalizing epistemology is strongly supportive to the scientific program writ-large, as his later work clearly states (e.g., Quine 1995). His view of science is "traditional" in many respects, and he highly regarded the fruits of scientific labors that followed a naturalism he strongly endorsed, which offered the best approximation of *objective, true* knowledge. I note his attitude because Quine wears two personae. The first is conferred by the power of his critique of the positivist agenda. That critique was within the same tradition of logical analysis that spawned the Vienna School, and Quine, trained as a logician, was very much a member of that tradition. Second, he was an influential interlocutor of Carnap (as well as other positivists) and, more importantly, he was one firmly committed to science as a form of knowledge; hence his naturalism. The other identification, one he would wear with discomfort, refers to the product of that critique, which set the course for postpositivism for the next half-century. That trajectory followed a radical social constructivist path from which he would distance himself (much as Kuhn professed hostility for the Kuhnians, who carried his work to ends for which he had not set course).

These caveats are necessary in order to place in perspective Quine's achievement, both for what he intended and for what he did not. After all, Quine's conclusions, *in toto*, were devastating to any normative account of science. Simply, the constructivism promoted by postpositivist historians and sociologists rests on the relativism Quine afforded them. If foundations were disassembled, what remained other than convention, consistency, and consensus? Forgetting that Quine embraced scientific reality as defined by a naturalized epistemology, more specifically, physical science, the irony is self-apparent.

Those who opposed Quine's followers by embracing a metaphysical realism and an earlier positivist normative philosophy of truth seeking were deeply disturbed by these developments. They held that reality would be discovered and increasingly defined as determined by what was *really* there. They asked, appropriately, how does science ever proceed without testing itself against something "real?" Without assuming some asymptotic approach towards some limit, in this case, the reality of the natural world, what does science *do*? More, by what criteria can "progress" be ascertained? Quine's naturalistic stance would rest itself on the pragmatic results of scientific scrutiny. That pragmatism would not satisfy those seeking more formal explanations, and so the issues he identified continue to provoke debate about realism and objectivity. On the other side of the aisle, historians and sociologists of science largely bypassed the philosophical realism/antirealism debates and focused instead on the methods and instruments that confer *objective* accounts. An emphasis was placed on "extra-curricular" influences, which were described under the banner of *constructivism*. The identification of such factors and their function in the production of scientific knowledge increasingly dominated the science studies literature in the 1970s and 1980s. While continuing to suffer from unresolved difficulties of definition (McMullin 1988; Hacking 1999; Kukla 2000), constructivism then served as the conceptual battleground of the ensuing "warfare" during the 1990s over how to define science and its place in society.

The Constructivist Challenge

When a logical structure for scientific discovery and verification seemed elusive at best (philosophy), and the history of science seemed similarly marked by nonprogressive, nonrational models of growth, students of science paid closer attention to the social variables that might account for scientific practice. Following Kuhn, the organizing question in science studies became how, and to what extent, aesthetic judgments, deep cultural and gender values, and social, economic, and political pressures combine to influence scientific rationality and method. A new ethos took hold: if positivist tenets became inadequate for such explanation, then an accounting of so-called "external" factors would supply a more comprehensive understanding of how science operates. And here, in the lacuna left by positivism's retreat, constructivism expanded its claims.

In the science studies community, constructivism "stands at the confluence of two streams in the history of sociology: the sociology of knowledge and the sociology of science" (Woolgar 1988a; Ashmore 1989; Kukla 2000, 7). The first domain, shaped by Marx, Carl Mannheim, and Emil Durkheim, assigns a causal role to social factors in forming individual belief, yet each exempted science and mathematics from social or cultural influences (7). As to sociology of science, the discipline experienced a radical change of course, shifting from Mertonian descriptions of science as a social institution to a post-Kuhnian "sociology of scientific knowledge" (SSK), where constructivism of various stripe appear in full flower.

With no prescribed method of inquiry or resting place for its truth claims, the dominant sociology of science depicted science immersed in, rather than riding above the needs of, the tribulations and power politics of science's supporting culture, lurching forward by rules not rigidly formalized through logical analysis. And because of the tight intercontextualization of scientific practice within a complex matrix of philosophical, historical, and cultural contingencies, these students of science argued that we cannot expect that science should possess a singular universal and prescribed method. From this general orientation, a spectrum of critiques stretches from the "strong" program of SSK (Barnes 1985; Bloor 1991) to various conjugates, for example, the empirical-relativist school (Collins and Pinch 1982; Collins 1992; Pinch 1986), ethno-methodological studies (Lynch 1985), actor-network theory (Latour 1987; Callon 1995), feminist epistemology (Haraway 1989a, 1989b; Harding 1986), and symbolic interactionism (Fujimura 1992). These various approaches are held together through a family resemblance that demands a circumspect assessment of how to situate science's appropriate intellectual claims in the context of sociological determinants.

A reasonable definition of constructivism begins with "X is said to be constructed if it's produced by intentional human activity" (Kukla 2000, 3), and thus human artifacts or social activities are easily recognized as products of human invention. But are scientific concepts constructed? Not according to the doctrine of natural kinds, whereby conceptual schemes carve nature at its preexisting joints, and these are then seen as discovered, not invented.[13] But this simply gainsays the success of the conceptual structure erected, the success, if you will, of the construction we call a theory, fact, concept, and so forth. At least that is the argument, which should be understood with certain distinctions, at times conflated:

Constructivists may argue a *metaphysical* thesis about the facts describing the world in which we live; an *epistemological* thesis concerning what we can know about the world; and a *semantic* thesis regarding what we can say about the world (Kukla 2000, 4). Some confusion exists in the literature as to what these various positions maintain, and, more pointedly, on whether constructivism is invariably associated with relativism, whether ontological or epistemological (4–6). The spectrum of alliances on this problem represents one axis of the constructivist controversy.

The second aspect of the debate concerns the extent to which one can make a constructivist argument in science. In some sense, constructivism in a social and political context is self-evident: certainly, science is pursued and supported for social reasons and technological gain. But the question looms as to what extent these forces *determine* scientific findings and theory construction.[14] Science is, in a trivial sense, "social." In other words, it is a human activity that draws upon all those elements of our culture that support its enterprise. This is hardly contentious in itself, but while describing the intercontextualization of science and its supporting culture seems innocent enough, such descriptions have generated heated debate when the arguments have followed a theoretical continuum that appears to conclude in radical deconstruction, the end point of which leaves science reduced to politics and in which an insidious relativism reigns. The ire of scientists (and here I am referring to those who regard science as an objective enterprise as commonly understood) targets the assertion that theoretical formulations are heavily determined by ideological orientations, whether political or sexual (e.g., Haraway, 1989a, 1989b; Harding 1986), or that scientific truth claims are no more than a rhetorical enterprise in which persuasion is used to overwhelm the opposition (Latour and Woolgar 1979). On this latter view, the pursuit of knowledge seems to command interest only as a process in which scientists are regarded as pitted against one another in an "agonist field," locked into a constant trial of rhetorical strength.

Not surprisingly, answers to the seemingly innocent sociological question, "How is truth erected or arrived at?" has, on the one hand, been interpreted by certain critics as a circumscribed, neutral description of the pathways governing scientific discourse (e.g., Collins and Pinch 1982; Collins 1992; Latour 1987), and, by others as a relativist assault on the scientific enterprise (e.g., Gross and Levitt 1994; Gross et al. 1996). To be sure, historians (e.g., Golinski 2005), philosophers (e.g., Hacking

1999; Kukla 2000), and sociologists (e.g., Cole 1992) have actively debated the merits of such constructivist case studies. Yet, as discussed below, the more extreme proponents successfully polarized the science studies community over the general question of how much of scientific truth claims result from a process of *discovery* of nature (as a realist posited), as opposed to those claims arising from a *construction* of facts and theories derived not only from apparent sense data, but also influenced (to varying degrees) by incipient cultural values and contingent historical, political, and economic developments.

Historically, current constructivist arguments take off directly from the positivist program. In the attempt to characterize the production of scientific knowledge, the status of objectivity is obviously central. In general, the constructivist challenges the realist's notion of strict objectivity and the independent existence that realism pursues. From that position, the fact/value entanglement becomes highly convoluted. At one end of the spectrum, the adherence to a strict objectivity as the basis of scientific discourse sets a certain array of governing positivist values. Indeed, the vanishing subject and the view from nowhere idealize this position. In contrast, those advocating a strong constructivist position recognize a host of social and psychological values that play on scientific discourse in ways that are highly threatening to an objectivity that is tuned to a reality simply there to behold or discover. This sociologically informed orientation regards the structure and practice of scientific institutions, and the wider political arena in which they function, as partaking in the *creation* of objectivity and its various judgments. Whether at the lowest level of fact gathering or at the last stages of theory development, the play of unacknowledged conceptual, linguistic, and psychological forces are deemed important in understanding the process by which science makes its truth claims.

In the setting of the constructivist assessment of scientific practice, positivist ideals seem quaint and naïve. Science, in fact, has no circumscribed boundaries, either conceptually (dipping into the reservoirs of various interpretative genres) or socially, as the laboratory gradually spreads into legislative halls, poll booths, newsrooms, courts, schools, and churches. Most importantly, reality becomes a "construction," albeit a highly successful one, but hardly *final* or *definable* independent of human categories with all of their irredeemable distortions and arbitrary organization. This disputed interpretation about what science *does* and *what* it captures builds on the philosophical shift that cascaded with the fall of positivism. Certainly, those developments opened the door for science studies to reconfigure long-held beliefs about the character of an epistemology riding above the confusion of human values.

When the study of nature and the study of society were seen as inexorably linked—not only interwoven in a trivial social sense, but locked together at their deepest roots—a radically novel picture of science emerged. It makes no sense, on this new view, to speak of nature (as science examines it) and culture (as historians, philosophers, or sociologists practice their studies) as independent domains. Needless to say, a fundamental debate ensued in the wake of this attack on old precepts, for the very conception of science had been radically challenged:

> [A]n anthropology of knowledge remains possible; but instead of being an explanatory and unifying meta-theory, it becomes the locus of dialogue between contradictory conceptual frameworks that determine different modes of defining what makes a fact a fact, different theories and different criteria of relevance. Even though criteria of truth can function in each of these frameworks, no single criterion traverses all of them. In terms of our own discussion, even though each game has its rules, there is no unique rule for playing with the games. (Atlan 1993, 370)

What heretofore was an inquiry into the social organization and practice of science based on some normative view of scientific discovery has evolved to encompass a sociological account of how scientific knowledge is generated and how its validity is established.[15] In other words, sociology has taken on epistemological concerns of *what* science describes and *how* it does so.

Beginning in the 1970s, sociologists embracing this constructivist orientation offered an alternative formulation to an "internalist" interpretation of science. This older attitude argued that scientific practice grows from local, immanent concerns, that it is subject to and governed by rational discourse, and that the world it examines may be discerned objectively by the scientific hypothetico-deductive method (Hempel 1966). The project to comprehend nature was thus viewed as essentially logical and self-sufficient. This model of "Science as Rational Knowledge" (Callon 1995) imposed severe constraints on the social organization of scientific work, and implicitly relied on the realist view of nature. Scientific practice became, fundamentally, a normative exercise, where social influences are minimized in the pursuit of truth.

The spectrum of alternatives to the "rational models" of scientific progress has embraced a strong sociological orientation and has appropriately been labeled a model of "Science as Sociocultural Practice" (Callon 1995). To this school Andrew Pickering ascribes no less than "a new approach to thinking about science . . . insist[ing] that science was interestingly and constitutively social all the way into its technical core: scientific knowledge itself had to be understood as a social product" (Pickering 1992, 1), and perhaps ironically aping its subject, the sociological approach is "determinedly empirical and naturalistic" (1). That argument, in its broadest interpretation, claims that science cannot be segregated from two levels of support: (a) the subtle and complex cognitive and linguistic infrastructure in which scientific practice is articulated, and (b) the economic and political forces that support the myriad activities of science. In fact, science is heavily indebted to these influences and thus objectivity, the métier of scientific truth claims, is seen as framed by various factors. Studies from this school have shown that we cannot understand "what is" (i.e., the reality as described by science) independently from how that reality is examined or produced in the laboratory. Thus the constructivists align themselves with the antirealists by arguing that knowledge formation is "filtered" through various conceptual and cognitive sieves, so that the ontological claims become hopelessly conflated with the epistemology employed to make truth claims. They discard positivist objectivity (arrived at by transcendental, timeless norms) and substitute pragmatic, local-realist demands. For these constructivists, objectivity is, in the end, a negotiated agreement among interested parties and holds no singular conceptual ideal attribute that might characterize it (Megill 1994). Accordingly, facts are laden with both declared and unannounced values, and, correspondingly, scientific methods are pragmatic and resistant to formalization, so that no "system" could comprehensively (or fairly) describe scientific method (Feyerabend 1975) or theory development (Kuhn 1962, 1970).

Polanyi's emphasis on the cognitive complexity, and irreducibility of scientific thinking laid the groundwork for Feyerabend's more radical attack on a singular scientific method:

> My intention is not to replace one set of general rules by another such set: My intention is, rather, to convince the reader that *all methodologies, even the most obvious ones, have their limits*. The best way to show this is to demonstrate the limits, even the irrationality of some rules, which she, or he, is likely to regard as basic. (Feyerabend 1975, 32; emphasis in original)

Feyerabend opposed a formalistic schema of rationality that followed some set of rules in an algorithmic or procedurally structured manner that, while conceived as necessary, universal, and atemporal, was clearly not (Farrell 2003). Indeed, simply by looking at scientific practice he found it obvious that merely understanding the rules, and even following them, hardly yielded scientific truth. Instead, Feyerabend offered what Robert Farrell calls "tightrope-walking rationality," strung between theoretical/abstract traditions and empirical ones. Inherent in each are a set of values that structure a theory of reason characterized as the context-balancing of inherently complex, even competing values (Farrell 2003, 188ff.). Within the theoretical/abstract tradition, the value of comprehensiveness is associated with such concepts as simplicity, generality, explanatory power, and consistency. The empirical tradition also brings empirical accuracy to play. To this, Feyerabend adds teastability and fecundity as completing a quadrivalent set of values to account for rational inquiry (Farrell 2003, 203). Indeed, there is no distinction between the kinds of rationality applied in different sectors:

> Where distinctions can be made is in the interpretation and weighing of the values . . . [and] differential interpretation and weighing of values is a ubiquitous affair which just as much distinguishes physics from biology, or nineteenth-century physics from twentieth-century physics as it distinguishes science from commonsense. (203)

Note that Feyerabend's ensemble of values are *values*, not rules, and thus he revealed a false conceit: any notions of methodological orthodoxy that might follow from a formula of inquiry had been dispelled. Perhaps "anything goes" (the absence of any rule structure) was hyperbolic, but he got everyone's attention, and rightly so. For our purposes, he moved the discussion from formalized rules to pragmatic application of values. This was a decisive shift.

The dominant theme in current science studies literature adopts a strategy that seeks to put the realism question to the side[16] to ground scientific belief in the notion of "reasonable practice." By focusing on *practice*, these scholars draw scientific truth from local realities. Such studies have focused inquiry on the laboratory, both the site of experimentation proper and its broader reaches, to describe the setting from which facts emerge.[17] We must distinguish local case studies from a second set of perhaps more traditional sociological approaches, namely, discussing how such knowledge is propagated (e.g., Rouse 1987, 125) and offering descriptions that focus upon institutional structures, political and economic factors directing research, as well as "cultural" and training practices of professionals and scientists in training. These two modes of sociological inquiry, one peering closely at laboratory practices and the other at the institution generating scientific knowledge, are linked, but the first focuses on the problem of the autonomous knower, which is our primary consideration here.[18]

The next chapter provides further details about how reason has been recast in the postpositivist era. No longer does the theorist's "glimpse of some aspect of the true structure of reality" epitomize the rationality of science. Rather scientific rationality involves a process by which the skilled practitioner coaxes usable observations to craft theories and "dynamically works out a more or less stable but always evolving accommodation between the provisional results of those two enterprises" (Brandom 2004, 4). This pragmatic orientation for understanding scientific thinking as first and foremost grounded in practical laboratory practice regards the core importance of science to be in its predictive power, in its empirico-logical praxis of constructing reality, and finally in its demonstrable ability to manipulate nature for technological applications. Hardly anyone would dispute the practical success of the scientific enterprise, but furious polemics arise when the discussion shifts to how science's theoretical claims are grounded, and, perhaps more importantly, how and on what basis to apply scientific knowledge to the social world. To these matters we now turn.

4

The Science Wars

It is doubtless impossible to approach any human problem with a mind free from bias. The way in which questions are put, the points of view assumed, presuppose a relativity of interest; all characteristics imply values, and every objective description, so called, implies an ethical background. Rather than attempt to conceal principles more or less definitely implied, it is better to state them openly, at the beginning.

Simone de Beauvoir, *The Second Sex*

If knowledge is social, if language is delimiting, if historical and social factors mold the scientific enterprise, what in the meeting of the "mind" and "nature" defines the real? Or more modestly, what are the epistemological boundaries of science in the quest of the real? The answers that have emerged in the past four decades present interpretations that, while different in form, converge on a "failure": science possesses no essentialist definition, which, at least in part arises from the collapse of rigid distinctions between facts and values. Some singular idealized objectivity and neutrality, long contested and now defeated, fails to satisfy the close examination of science's history, philosophy, and sociology. Instead, facts are understood as contextualized in theory or models, where they acquire meaning. Beyond this dimension, a larger scaffold of social, linguistic, and historical values frame, to varying extents, those models. Accordingly, facts fit into constructs through the application of explicit and implicit values—both intracurricular and extracurricular to a scientific grammar as normally construed.

Since a simple positivism by which science might be understood has become inadequate to describe the medley of scientific practices and the mosaic of its logic, the focus of attention has shifted from the philosopher's analytical attempts to characterize scientific discovery and verification to one tilted prominently towards analyses that take into account the sociology of scientific inquiry. Science studies has moved beyond an idealized vision of scientists cloistered together in a community exclusively dedicated to the unprejudiced pursuit of truth, toward a conception of science as including social pursuits not qualitatively different from those of other social constellations. On this view, science is not simply governed by its own inner logic. Aesthetic judgments, deep cultural values, and social, economic, and political pressures combine to form a complex matrix in which scientific rationality and method are influenced by extracurricular elements. With no foundations, the logic of science becomes part of a swaying edifice of values and perspectives. The so-called disunity of science then is more than a topology of different conceptual schemes, where various parts of nature are described with tenuous connections to other scientific disciplines (Dupré 1993; Galison and Stump 1996). Rather, disunity characterizes *science's own reason*, the epistemology of which its boundaries and inner structure, exhibit as an "open architecture."

Without uniform methodological characteristics, inquiry is pushed in one direction or another in response to diverse demands that require interpretation of different kinds, which then invoke different kinds of judgments and values to adjudicate knowledge. At Quine's tribunal of experience (1953a/1961/1980), those interpretations follow broadly defined epistemological rules. Cohesive accounts emerge, but the picture of science's reasons then looks, at least from this vantage, as a scaffold with girders waving to and fro—standing tall as they accommodate the winds that buffet them. A more conservative depiction, one which perhaps nostalgically peers at a lost innocence, draws the blueprints of the manifold with more precision to contain the allowed limits, to restrict poorly defined latitudes, and to impose fewer degrees of freedom. Note, however, the picture is essentially the same, and the basic role of values configuring the structure remain.

To capture science's discourse, we peer at one picture and then the other to finally behold a diptych: science in its local (or near) dimension is governed by a set of rules, the logic of which appears generally uncontested. Here, objectivity reigns satisfactorily to reveal "the real."

This is the science of the practicing scientist, whose work could not abide the ruthless, reflexive exercises that the philosopher enjoys. The contrasting picture offers a different perspective, because the student of science assumes her position at a more distant vantage. From there, science's reason appears influenced by contributions beyond some predictable, ordered reason. These include various social and institutional influences, external political demands, psychological elements such as aesthetic imagination, financial and professional rewards, and so on. Steadfastly hooked to nature, this multidimensional reason creates as it discovers reality, presenting a composite between "nature" and "mind" that continuously bumps against the limits of each. *How* that picture has developed is the study of scientific achievement and the evolution of human consciousness which ultimately administers it. The rules of governance have exhibited no resting definition or steadfast rules, so the activity called "science" constantly requires critical evaluation to reveal its inner workings, its faults and weaknesses, and the creative machinations, which produce the scientific product. As a *philosophical* pursuit, this view of science places it well within its parents' fold. Indeed, science studies arrived at a place quite different from its predecessors and, in that movement, encountered resistance. The controversies that followed Kuhn's *Structure* eventually led to the Science Wars of the 1990s.

Battles in the Night

As the previous chapter described, the fundamental issue preoccupying science studies has been the degree to which scientific findings—from theory to elemental fact—are constructed (Zammito 2004). This issue spanned the axis of the sociology of knowledge, extending from erudite philosophical analyses to the political standing of science and its role in adjudicating social questions. Essentialists maintain the possibility and analytic advantage of identifying the unique and invariant qualities that set science apart from other occupations and thus explain its singular achievements. Thus far, I have referred to this group as the "defenders of science," by which I mean they still hold to notions that follow a positivist philosophy of science. Constructivists, along a continuum of critiques, basically deny any such demarcation and instead maintain that science, like other intellectual disciplines, is contextually contingent and driven by the pragmatic interests of its supporting political culture. This latter group dominates the collective of disciplines comprising science studies, but it divides into two branches.

The first branch assumes a more radical postmodern position, where relativism and ideological bias, as opposed to objectivity and truth, frame their appraisal of science. Allowing for its particular object of study (the natural world), the radical critics not only see the demarcations that had previously distinguished science from other conceptual pursuits as pulling science back from its distinctive status to dwell among kindred faculties, but regard science's own logic as governed by strong social factors. The second, more conservative group embraces a pragmatic, circumscribed attitude to argue that science requires a multifocal appreciation of its methods and truth claims. Furthermore, while the canon of scientific method has been replaced with a multiplicity of strategies, and scientific progress follows a pragmatic course (not some prescribed logical path of inquiry), this group, by and large, still allows for a large measure of reason and objectivity to reign in the laboratory. While both "radicals" and "conservatives" might agree that their respective criticisms could hardly diminish science's technical accomplishments, their descriptions of the relationship of scientific reason and other kinds of knowing markedly differ.[1] The polemics described in this chapter pit the radicals against the defenders, and, in some sense, the conservative constructivists are caught in the middle. They are far from a discarded positivism, but they seem to implicitly agree that the thrust of their work is to understand why knowledge acquisition in the sciences is special. This "score card" of the players configured in a dramatic contest admittedly caricatures the dispute, but as we proceed to describe the Science Wars, we must distinguish the radical critics from those who did not join their ranks.

Although largely an academic affair, the repercussions of the controversy reached into many corners—the judiciary, public policy, education, and beyond. Indeed, the scholarly squabble revealed, again, how the rationality governing science is contestable, and hotly so. While the dust has settled in some quarters, scars remain and many wounds are still unhealed. The argument, in its simplest expression, concerned the character of scientific truth claims, where those holding to a scientific realism fought a defensive battle against those who regarded scientific knowledge as exhibiting various degrees of construction to account for a depiction of reality. Those battles were specifically fought over the character of objectivity and how reality might be described. In another dimension, the argument centered on the relative roles of epistemic and nonepistemic values in the governing of science.

The challenge to ponder the claims of the scientific vision, not only its characterization of nature but also its portrayal of society and human kind, spilled over the borders of science. To be sure, this divisive controversy over the sociology of knowledge (i.e., the values of science, the relation of science to its supporting culture, the nature of scientific reasoning, etc.) was perceived as vital to intellectual inquiry, generally. Consequently, the debate carried over to other disciplines. In similar voice, historians (Fay, Pomper, and Vann 1998) and literary critics (Adams and Searle 1986) pitched themselves in battles over historiographical integrity and deconstruction of texts, respectively. Thus with the rise of this new criticism, intellectual territories were marked and divided not necessarily by different academic traditions and subjects of inquiry, but rather by how one regarded the character of knowledge, rationality, and objectivity. The crisis, begun in philosophy, quickly expanded to history and sociology of science.

Some contemporary sociologists of the constructivist camp seized the mantle of antirealism in their sociology of knowledge analyses and extended it beyond what most antirealist philosophers probably intended. From that radical perspective, relativism enlarged its domain to the point that the scientific community, which saw a threat to its basic precepts, sent out an alarm (Holton 1993; Gross and Levitt 1994; Gross et al. 1996; Ross 1996; Barnes, Bloor, and Henry 1996; Koertge 1998). And just as there were diverse perspectives offered by social constructivists of various stripe, the rebuttals were similarly slanted towards different agendas, whether the argument was poised in historical and cultural (e.g., Holton 1993), sociological (e.g., Cole 1992), or philosophical terms (e.g., Laudan 1990a).

It is too early to judge the full ramifications of this debate and assess the proper application of the various brands of social constructivism, that is, the extent of social influence in determining communal scientific knowledge and the constraints of the empirical world on cognitive relativism. But what remains most appealing about science studies, if a generalization is to be made, is the barbed critique of what Harry Collins has called the "ethnocentrism of the present," that is, the feeling that "now" (as opposed to the past) we have finally achieved scientific "maturity" and have a method and a worldview that approaches some Final Theory (see Weinberg 1992 for an example of such a theory). The skepticism regarding universal and unchallenged modes of assessing scientific statements, or for distinguishing between intrinsically scientific and nonscientific research programs, has been forcefully reiterated by recent science studies, which *in toto* seem to share doubt that one singular, rational scientific methodology has finally arrived.

This critical position, one supported by both philosophical and historical analyses, hardly seems novel. After all, skepticism forms the very foundation of research practice, an ethos that has ruled science since its birth. Recent sociological studies have reiterated that message but perhaps have formulated it in a more complex and more textured manner than previously appreciated (Gieryn 1995). What seems new are the angry rebuttals from science defenders, who perceive that constructivist criticism has hoisted the banner of relativism over science (Gross and Levitt 1994; Levitt 1999). They regard the definition of science as being at stake, as well as the definition of truth itself. And they ask what is real if so much is contingent. Led by Paul Gross and Norman Levitt, these science defenders have often misrepresented and distorted the constructivist position (e.g., Wendling 1996; Tauber 2000). Putting aside distinctions between "weak" and "strong" forms of social constructivism and the various permutations adopted along realist/antirealist, correspondence/coherence, and foundationalist/antifoundationalist lines, the constructivist simply holds that some empirical truths are partly constituted by cultural beliefs. Reality includes the entire phenomenal world, both the empirical and the social, and, in the end, how we classify reality is a social matter. The issue then becomes, "which *aspects* of reality are social, *the degree to which* they are social, and *the extent to which* they can be changed socially" (Wendling 1996, 425).

So the overarching question to ask of the social constructivist is to what degree science practice, as socially determined, allows its *conceptual* work to be constructed from extracurricular sources (i.e., those that are construed as outside scientific thinking as generally understood from an internalist perspective)? In other words, to what degree do scientific theories, hypotheses, even facts, reflect the influence of societal mores, expectations, and underlying cultural structure, as opposed to the expressions of scientific experience that are dependent on a rhetoric of common language?[2]

The armies joined in the Science Wars displayed vociferous appetites for the conflict. The science studies scholars claimed that "the history, philosophy, and sociology of science should not be entrusted to practicing scientists" (Lynch 1993, 268). Accordingly, while scientists offered crucial self-portraits and eye witness accounts, these were not generally regarded as sufficiently sophisticated or detached to achieve an analysis of discovery, theory formation, methodology, scientific practice, and so on. These latter efforts required a reflexive stance, a perspectival "distance." On the other side, defenders of science decried the postmodern tenets held by their opponents, whose attacks on the citadel of Reason and Truth were tantamount to dismantling the greatest cultural achievement of Western societies. Thus the alleged distance—the moat, if you will—separating the science studies scholar and scientist was easily crossed, and when the walls of the laboratory were breached, hand-to-hand combat ensued (Gross et al. 1996; Ross 1996; Haack 1998; Brown 2001). Little useful interchange occurred. In a sense, the critics came to a party to which they were not invited, and once there, they would not leave.

The discussion on both sides was dominated by rhetorical hyperbole, for example: "There is no goddess, Truth, of whom academics and researchers can regard themselves as priests or devotees" (Heal 1987–1988, 108). The defense responded with impassioned and sometimes strident rebuttal, for instance, "It is downright indecent for one who denigrates the importance or denies the possibility of honest inquiry to make his living as an academic" (Haack 1996, 60), so one would justifiably banish such "cultural garbage" propounded by academic "slobs" and their collective "gangs" (Bunge 1996, 110, 96, 97). Taking the sole proprietorship on honesty could not foster discussion, much less a resolution. The acrimony peaked in 1996 with the Sokal Hoax (Editors *Lingua Franca* 2000). Alan Sokal, a physicist, wrote what he thought constituted a parody of the radical opposition's voice and submitted the paper to a respected critical journal (*Social Text*), whose editors published it, ignorant of the ridiculing intent (Sokal 1996). Sokal showed all contestants the vulnerabilities of the more outlandish critiques and, at the same time, pointed to the need for some common ground for honest debate (Sokal 2008).[3]

The indictment was clear. The radical social constructivist stood accused of arguing a damning relativist position: according to radical constructivism, beyond the usurpation of common language and public categories to describe nature, the scientist manufactures theories from the social residua of his hidden prejudice and preformed social consciousness so that the conceptual product is only the social world portrayed in a different, albeit "natural," guise. (For examples of this view, see Collins

1992; Harding 1986; and Haraway 1989a, 1989b.) In other words, language is so imbued with the social construction of reality that scientists cannot legitimately separate scientific facts from their supporting cultural milieu. Further, some radical constructivists like Steve Woolgar (1988a, 1988b) extended the argument by contending that science uses a rationality designed for its own hegemonic ends and thus would enlist a radical self-reflexive sociology as part of a far-reaching ideological battle over the very nature of knowledge itself. Indeed, the social consequences would be revolutionary, possibly seismic, if the radicals succeeded in fatally undermining the testimony of objective science (Ashmore 1989). They might be easily dismissed as part of a more complex ideological program within which science—in the name of some higher moral vision—became an object to be dismantled. But even those more circumspect about the ability to define scientific inquiry and the rules that govern it, earned angry responses from those who regarded attacks on the "scientific method" as breaking ranks with rational discourse.

Rejoinders reprimand: science *is* successful, and to conflate the contextuality of knowledge with the compromise of its objectivity is to deny the obvious accomplishments of scientific method. No doubt scientific objectivity depends in part on its contextuality, but that is not to deny its proven strengths of verification, coherence, and predictability, even within the local context of scientific inquiry. So, while acknowledging that social factors do indeed play an important role, we must not lose sight of how science locks onto nature, offering means for manipulation and powerful prediction. Science offers no arbitrary description, and within its local domain, scientific investigation serves as an important limit to relativist thinking. Certainly, while scientists might choose among various research programs by what may be judged as "ideological" (e.g., Tauber and Sarkar 1992, 1993; Lewontin 1991), or "aesthetic," criteria (see Kuhn 1970 and the respective essays by Margolis, McAllister, and Sarkar in Tauber 1996b), or even as a consequence of trivial options dictated by disciplinary traditions or commitment to broader social concerns, those determinants hardly gainsay the work objectivity performs. (This position represents the basic opinion of the "conservative" constructivists identified above.)

Many aspects of the controversy deserve further scrutiny, but one issue in particular highlights the self-confident enthusiasm of some radical science critics. It is one thing for sociologists and historians of contemporary science to obtain facts and to attempt analysis by means that have proven so successful in the natural sciences. This may be a legitimate project, but a not-too-subtle inversion takes place when, rather than applying natural scientific method to social analysis, the historical or sociological study becomes integral to science itself, or at least is perceived as such (e.g., Söderqvist 1997 and critiqued by Tauber 1999b).[4] The boundary between science and its study then becomes hopelessly blurred: "[We] see the sociology of scientific knowledge as *part of the project of science itself*, an attempt to understand science in the idiom of science" (Barnes, Bloor, and Henry 1996, x; emphasis added).

To be sure, at one level, these critics asserted their methodological legitimacy as social scientists (i.e., using the "idiom" of science and thereby "honour[ing] science by imitation" [Barnes, Bloor, and Henry 1996, x]).[5] At another level, they did more than ape scientific method in the project of understanding science, for these sociologists maintained that they were engaged in the "project of science" itself, which must include the study of human organization and behavior. Science then, extends beyond the examination of the natural world to include culture studies and, reciprocally, the unique standing of the scientific endeavor is now open to other kinds of analysis that includes sociological or historical comment. If the net is cast wide enough, the entire enterprise becomes one piece. The structure of knowledge then has no boundaries or subdivisions, and history and sociology have extended their own imperialistic ambitions to include science itself within their own bailiwick. The implicit challenge is over the very definition of science, which no longer, from this point of view, can claim any autonomy.

If the traditional internalist understanding is invoked, scientists in the laboratory do science, whereas those who report on science do so from the outside (i.e., externalists), whether it is as sociologist, historian, or philosopher. But this designation breaks down when the notion of scientific knowledge and the modes by which it is obtained are perceived as essentially no different from other forms of human understanding. This perspective is now widely held in science studies. Some regard the issue as an instance of simple politics, for example, "Science itself is not SCIENTIFIC except in so far as it represents itself as such" (Woolgar 1988a, 107; emphasis in original). This position rests on what Larry Laudan refers to as an epistemological conceit: with the emergence and eventual triumph of the fallibilistic perspective in epistemology, it became generally accepted that science offers no apodictic certainty, and thus all scientific theories are potentially corrigible and subject to serious emendation. Once the foundations have been shaken, the straight line drawn between knowledge and opinion begins to meander. And without firm demarcation criteria, distinguishing science from nonscience becomes problematic and allows the relativist to demand new criteria or dispel the authority of the old standard. Within this fallibilistic framework, scientific belief then turns out to be a species of opinion (Laudan 1996, 213).[6] If one adopts this general orientation, the separation of observer from his or her object of study—whether science or history—may be easily blurred and the positivist program crumbles all the way down to its very foundation.

The Character of Reason

The arguments at the center of the Science Wars go well beyond characterizing science as a form of knowledge. Many perceive that the assault of contemporary science studies is not only on a positivist vision of science, but, more deeply, on the character of reason itself (Wollgar 1988a). If the rationality upon which science's enterprise is based becomes a focus of challenge, the argument has moved well beyond the traditional questions in philosophy of science. For example, such questions as, "To what extent may science make its truth claims?" or "By what criteria are we to judge its avowals of rationality and objectivity?" traditionally assumed reason's autonomy. The new critiques do not, and rationality itself has no essential structure unique to scientific practice:

> Science may be complex, they say, but it is still "rational." Now the word "rational" can be used either as a collecting bag for a variety of procedures—this would be its nominalist interpretation—or it describes a general feature found in every single scientific action. I accept the first definition, but I reject the second. In the second case rationality is either defined in a narrow way that excludes, say, the arts; then it also excludes large sections of the sciences. Or it is defined in a way that lets all of science survive; then it also applies to love-making, comedy, and dogfights. There is no way of delimiting "science" by something stronger and more coherent than a list. (Feyerabend 1975, 246)

To recognize that science lurches forward by rules neither rigidly formal nor necessarily logical nor insular to its narrow interests forfeits an old conceit and instead embraces a more comprehensive understanding of this complex activity (Feyerabend 1975). Accordingly, such circumspect assessments of science's appropriate intellectual claims and the power of its vision (not only of nature, but of society and ourselves) enunciates a more honest appraisal. Pragmatism rules.

By defining scientific reason in a constructivist formulation, where the calculus of value explicitly figures in its operations, two schemas of reason, in dialectical interplay, present themselves—*discursive* and *social.*

(1) Discursive reason is formulated within a long philosophical tradition, in which rationality and advancement of science owe far less to a confident reliance on data, methods, and warrants than to the self-doubting Socratic "dialectic of interrogation" to which facts and theories are regularly subjected (Fisch 2006). Incapable (as a matter of logic) of objectively confirming their efforts, let alone of *proving* them, the scientist can, in principle, boast no more than to have prudently subjected her work to the most thorough tests available. That knowledge is incomplete and must be scrutinized through the lens of skepticism remain the key precepts of critical investigation of all kinds.

This epistemology serves science as it did philosophy from its earliest awakening. Socrates specifically contrasted such reasoning to revelation and opinion. By endless interrogation, he drove his interlocutors to face their complacent assumptions and lazy beliefs. He thus established the basic demand of philosophical inquiry. Fallibility is the linchpin of the entire enterprise, for the body of knowledge is assumed to be incomplete, if not in error (Popper 1963, 228ff.).[7] The perfectionism of endless critique provides the scientist with the basic value of inquiry, a value which binds science to its philosophical antecedents. Doubt and skepticism remain the cardinal virtues of scientific theory as well as underlying its various modes of proof.

Derived from this self-critical foundation, science developed values that seek to legitimate interpretation by parsimony, coherence, and predictive capacities. And success is assessed by an on-going rational criticism:

> Entertaining a doubt adds up to little more than applying a question mark, or raising one's eyebrows; serious criticism, by contrast, requires fashioning an argument. To doubt is to suspect something might be amiss, to criticize is to *argue* that it is. Skeptical discourse requires a supply of interrogatives, critical discourse requires rich background knowledge and a developed logic of problem-seeking and solving. Criticism necessarily presupposes doubt, but is also a necessary prerequisite for positive action. In the face of

> suspected imperfection the first step toward improvement will always be critical. Hence the term "constructive skepticism." (Fisch 2006)

Rationality on this view becomes a category of action, a means to expose and solve problems, and a method by which inquiry might gauge its success or failure as determined by a broader set of goals and standards. This instrumental quality of rationality opens the inquiry's frame of reference. The local problem is set in a context that itself has an orientation, a set of larger issues that confer direction to the more local investigation. In a sense, the critical position is obtained by standing outside the local framework, with persistent reference to local strictures.

This description, which appears imminently credible, holds sway among those who hope to save science's own logic in terms that follow some general rules of operation. Indeed, even Feyerabend's critique invokes a mode of understanding rationality by describing its guiding values, and that effort suggests not only that he was not as radical as he liked to posture (Farrell 2003), but that his tempered argument offered a template in which to understand the reason of science. After all, science works, and we would do well to understand why! The scientific enterprise is committed to a kind of verification not found (and usually unattainable) in other domains of knowledge. Despite the obvious successes that rely on this claim, objectivity as understood in the scientific context rests on a complex philosophical foundation that remains contentious. This is a highly complex issue, but in its simplest formulation, the argument concerns how criteria are chosen for successful prediction and verification.[8]

(2) In contrast to "discursive reason," "social reason" has laid claim to a constellation of scientific practices as *constitutive* to scientific reasoning, whose intellectual dimensions cannot be separated from the full context of science in action, with all that such a conception entails. Science as "social" becomes a trivial observation in the sense that science as an institution is *social*, as are all collective human activities. But to call science "social" as a system of knowledge, which is *socially constructed*, challenges standards of objectivity and scientific method that have resounded for generations.

As already discussed, social constructivism has varying meanings and applications, but all confess to a vision of scientific practice as hardly value-neutral. Thus, instead of riding above the tribulations of social intercourse, scientists are seen as engaged in projects, which draw from many linguistic and conceptual (cultural, historical, philosophical, sociopolitical) reservoirs, only one of which is "nature." Indeed, reality only emerges as a function of how it is revealed by the human mind and its apparatuses—technical, conceptual, *and* social.

On an historicist account, science today, relative to yesterday, depicts a different view of the world, and in the recounting of how science at any given stage made its truth claims, we witness how criteria of verisimilitude have undergone change, and, naturally, the scientific theories arising from those standards. "Truth" then is a moving target, a category that helps organize the investigative endeavor, but no criteria avail themselves for approximating how close current theories are to some final vision. If this is relativism, then it is only interesting as a sociological depiction, namely, reality looks different now, as opposed to then (as well as different from here as opposed to there). Accepting such a relativist position does not allow the dismissal of scientific knowledge as somehow an arbitrary construction or that reality is *only* a construct. If the relativist is confident to go up in an airplane or subject himself to open heart surgery, then debunking science as a construct (and thus somehow arbitrary) cannot be his game. He cannot be saying that science provides as "truthful" a picture of nature in 2008 as it did in 1608. Rather, he holds to a picture of truth that is the best consensual approximation to some unknown ideal and the most pragmatically useful instrument for various ventures, again as determined by the group's own notions of human flourishing (Latour 1999). Accordingly, each era holds reality within its own cultural moment, and that scheme changes with time.

If change and uncertainty are implicit to what we ascertain as "the real," then no resting place exists, no final theory, and of course, no complacency. This attitude originates in the birth of philosophy, namely Socratic reason, whose interrogation is ceaseless. As we swing between discursive and social reason, we need not choose one over the other, for each refracts the scientific enterprise accurately from its own position. The trick is to find their accommodation to each other, and ultimately, their union. Yet a gulf seems to separate these underlying notions of reason: while each school might agree that science grows, the lesson to be learned is *how* it develops.

Discursive reason argues from an "internalist" formulation—scientific thinking follows some internal logic—while social reason maintains that such logic is not only impossible to demonstrate, but also fails to fulfill its *own* aspirations. Ironically, these assertions take discursive reason to its own logical end, where an open-ended process can only provide another question driven by the presumed fallibility of the position currently held. Indeed, I do not regard the constructivist (even relativist) position so interesting as either an epistemological argument or even a metaphysical one, but rather it asserts a moral position about science and the values that govern it. Richard Rorty cogently makes the case:

> [T]here is no reason to praise scientists for being more "objective" or "logical" or "methodical" or "devoted to truth" than other people. But there is plenty of reason to praise the institutions they have developed and within which they work, and to use these as models for the rest of culture. For these institutions give concreteness and detail to the idea of "unforced agreement." Reference to such institutions fleshes out the idea of a "free and open encounter"—the sort of encounter in which truth cannot fail to win. On this view, to say that truth will win in such an encounter is not to make a metaphysical claim about the connection between human reason and the nature of things. It is merely to say that the best way to find out what to believe is to listen to as many suggestions and arguments as you can. (1991b, 39)[9]

The undermining of scientism as absolute or final knowledge reflects a deeper moral apprehension concerning the potential tyranny of a narrowly conceived rationality and the requirement for continued vigilant protection of open inquiry and free exchange of ideas (Sorell 1991). Again, this is the modernist project taken to its own internal conclusion. That interpretation interests me more than the argument about construction as an epistemological discourse because it begins to uncover other issues that the contentious fights over the social construction of knowledge hide. I am most interested in this moral dimension of the argument, one that is buried beneath the other debates.

Again building on Rorty's formulation, rationality is not only a commitment to a certain kind of methodical thinking, but also refers to being reasonable. Rational, in this latter sense:

> names a set of moral virtues: tolerance, respect for the opinions of those around one, willingness to listen, reliance on persuasion rather than force. These are the virtues which members of a civilized society must possess if the society is to endure. In this sense, "rational," the word means something more like "civilized" than like "methodical. . . ." On this construction, to be rational . . . eschews dogmatism, defensiveness, and righteous

> indignation. . . . My rejection of traditional notions of rationality can be summed up by saying that the only sense in which science is exemplary is that it is a model of human solidarity. (Rorty 1991b, 37, 39)

In "defending" science, some proponents have ironically placed themselves beyond the reach of liberal discourse. The deeper moral agenda of many critics is to show how value-free science is a conceit, and more, a distortion of its own methods. By opening science to such criticism, a wedge is placed for the kind of open dialogue Rorty advocated, one which lies at the heart of science as an exemplar of critical exchange.

Taking Stock

Much ink has been spilt on the Science War controversy, disagreements amply aired, epistemological and moral stakes clearly articulated. As the polemics have ebbed, it is time for an accounting. With the benefit of hindsight, we might well regard the polemics as no more than an academic civil dispute—overblown, strident, misconceived, hysterical; indeed, some argued that declaring a "war" not only misapplied the term, but contributed to the ferocity of the polemics (Shapin 2001). But the wounds are still fresh because the diatribes were understood as having profound significance far beyond the claims of science to include the foundations of ethics, the rational basis of public policy, the basis of personal beliefs, and so on. The issues raised in the Science Wars obviously call for attention. To be sure, philosophers have long grappled with the claims of naturalism versus constructivism, realism versus antirealism, positivism versus relativism. While these epistemological questions originated in ancient philosophy, the grounds for the current debate shifted from the nineteenth-century romantic attack on positivism to a more subtle argument about the strictures of language, the nature of objectivity, and the limits of rationality that mediate scientific knowledge. The basic problems concerned the degree of objectivity and neutrality science employed in describing reality, the insularity of scientific method, the social character of truth, and, most generally, the specter of relativism (Hollis and Lukes 1982).

Resolution of these issues, not surprisingly, remains elusive, but a treaty of sorts, or at least a ceasefire, seems to be in effect. The disputants of the more narrow Science Wars have made an uneasy peace, now that the sociologists' claims have been heard and the science studies position better understood (Labinger and Collins 2001). Clarifications of the issues have several sources, but Peter Dear, at least for me, resolved the matter most decidedly with his invention of the term "epistemography" to describe sociology of scientific knowledge, which distinguishes knowledge (or truth) from how knowledge (or truth) is made:

> Epistemography is the endeavor that attempts to investigate science "in the field" . . . asking . . . What counts as scientific knowledge? How is that knowledge made and certified? In what ways is it valued? . . . It designates an enterprise centrally concerned with developing an empirical understanding of scientific knowledge, in contrast to epistemology, which is a prescriptive study of how knowledge can or should be made. (Dear 2001, 130–31)

Epistemography requires an objective attitude, which simply means that the social scientist can make no commitments as to whether scientific knowledge is true or false. Accordingly, from the sociologists' perspective, when confronting either the realist or the relativist, the issue of scientific success is hardly addressed, inasmuch as they largely remain agnostic about the respective merits of various methodological and evaluative scientific strategies (e.g., Latour 1987). Thus the matter is not even placed on the agenda, for sociologists are content with a description of science as simply one among other social activities.

Indeed, truth is not at issue; belief is. This distinction was lost and the position assumed by the epistemographer brought the charge of relativism upon Dear's agnosticism, to which he responded:

> People do not believe propositions to be true because those propositions are in fact true; instead, they believe things for various reasons that an epistemographer is interested to uncover. . . . The complete analytical divorce between truth and belief means that the epistemic relativism attaching to beliefs has no bearing on any kind of position that maintains that truth is somehow "relative." An epistemographer could maintain a belief in absolute truth and still be a methodological relativist, because the relativism only applies to understanding what particular groups of people believe and why they believe it. (Dear 2001, 131)

To study science as a form of knowledge reveals that its belief structure, beyond truth claims based on factual data, includes multiple layers of interpretation, various modes of social and political intercourse, and a complex array of interlocking belief systems that may or may not have the same epistemological standing as the original claim. This is not to say that science does not produce truth, but only that scientists try to make truth claims about the natural world, and science studies scholars try to characterize how scientists do so. Such a critic might well be a methodological relativist and yet still hold to a belief in absolute truth. Note that while the objects of study are different, a common understanding of reason is held by both sides. Each seeks objectivity and impartiality. That commonality serves my own attitude.

I do not contest that science has followed a unique intellectual heritage and developed distinctive methods, but the "separatists" fail to recognize a deeper stratum of shared inquiry and purpose. To discern that shared agenda requires appreciating the array of values governing the institution of science, whose various activities go well beyond the walls of the laboratory. It also calls for appreciating the notion of a shared reason, or at least reason as it functions in some interplay of its various forms.

On this view, the mission to critique science, interpret its development, and assist efforts made from within the scientific establishment in its own self-critical evaluations warrants science studies. After all, it seems self-evident that the ability to translate scientific discoveries and theories into wider conceptual and social contexts, where their significance might be more fully appreciated, requires ongoing critical appraisals. Science is only one system of investigation within that larger arena of human study of nature, persons, and society. As such, it has proven to be a crucial means of discovering our world and characterizing our relationship to it. But like any mode of study, it is subject to criticism, and in that critique, scientific method itself is scrutinized and thereby understood in its widest context. This is part of the humanist enterprise in which science originated and to which it continues to contribute. One might even say that self-critical scientists are themselves engaged in the philosophical project of "natural philosophy" in their own examination of methods and truth claims. This essentially philosophical self-evaluation represents a fundamental link between science and philosophy, and so we return to our original challenge of defining science in a humanistic context.

On a synthetic view, science's broadest philosophical agenda aligns itself with the humanist tradition, and while that alliance has suffered tensions, the larger liberal program must bring them again into close proximity. So, despite the obvious contrasts in their respective pursuits and methods, an abiding alliance between them is, from my point of view, natural and necessary: natural, because they are linked by their shared histories and kindred philosophies; necessary, because only their combined strength will sustain liberalism and assure their own welfare.

However, this alliance has been in crisis and its future remains unclear. Building on their technical expertise and cognitive separation, scientists have gone to great lengths to distinguish their methods from those characteristics of humanist study and humane experience. Not only are the objects of inquiry different, but the very "logic" of the laboratory putatively rests on different criteria of reason. Nevertheless, deeper affinities exist with science's intellectual precursors. Indeed, the separatist history is illustrative of a false dichotomy that has obscured the relationship of scientific thinking to its underlying philosophical sources. The question at hand is to what degree science remains natural philosophy, a branch of inquiry tied to other disciplines by a shared common rationality. Further, a shared commitment to the values instantiated by open inquiry and interpretation binds them. These values govern science and support the foundations of liberal society. Thus to understand science as a product of Western culture requires acknowledging its role in buttressing entrepreneurial freedom—free inquiry; unfettered judgment; uncompromised plurality of thought, which collectively casts science in its broadest context, one crucial to the vitality of Western societies.

So what conclusions might be drawn? First, the constructivist challenge has yet to play its full course, but one major result from contemporary studies is that it has demonstrated how science builds knowledge in an analogous fashion to other kinds of knowledge formation. While claiming a privileged epistemological standing, scientists might argue for a difference in degree, not in kind, of constructivist elements. That strictures exist is generally accepted, but controversy remains as to the degree to which social forces are allowed to determine scientific content and thus affect the criteria of what counts as a fact and its placement in some theory or model. If "mind" interposes itself, then hidden value, prejudice, and perspective linger as persistent factors. In short, can the mind directly know the world? Constructivists of various stripes have alerted their audience to the various degrees of constraint on direct knowledge (following Kant), and in doing so, they have profoundly challenged scientific realists.

Second, a reasonable alternative position has been offered, one which argues that scientific explanation follows a pragmatic course, and that, idealized truth claims of theory notwithstanding, science is driven by its practical concerns. Science asserts a powerful understanding of nature that cannot be denied if its claims are circumspect to its method of study and we are able to differentiate the validity of its range of assertions. Guided and prodded by its political environment, and implicitly structured by its underlying metaphysics, science remains a pragmatic enterprise resting on a constructive empiricism and the power of "local" descriptions (i.e., well-defined experimental systems) of reality. Taking that position, science's truth claims appear *provisionally* true. Note, "provisional" and "fallible" are not synonyms of "contingent." On this view, pragmatism assumes top billing as we seek a resting place between the two extremes of an untenable realism and a relativism that regards science as nothing more than a social construct.

In describing that resting place, we do well to return to the heuristic model developed by Lakatos to explain scientific change (1970) and adopt it to orient these epistemological issues. Let us regard the body of scientific knowledge as consisting of complex levels with an essentially stable core that remains unperturbed by the conceptual and social whirlwinds rushing around its more peripheral activities, the sites where new knowledge is being fashioned. These peripheral sites are more subject to various theoretical interpretations that include the influence of factors arising from the cognitive commitments and social setting of the research. But eventually, at some point in the trials of scientific theory construction, from its own theoretical vagaries and the uncertainties of obscure conceptual variables, the conceptual basis is "hardened" by further testing against nature. No longer subject to radical development, a research program matures, and its core content is protected and codified. In other words, how scientific theories emerge may well be examined as products of a particular scientific method, the political interests of the supporting culture, the historical contingencies of the setting, and so forth, but the evolution of that process does result in a *practical* achievement of effective manipulation of nature. On the basis of Newton's laws, we have landed on the moon . . . and returned. Beyond some social consensus, there is an objective, *qua* real result related to our manipulation of nature. That "product" constitutes an end point, regardless of whether or not the theory that explains it is true in any final sense. In other words, much "space" exists between scientific findings, which describe nature, and the theories, which serve to explain those findings. Indeed, an acknowledged hierarchy of validation applies to scientific knowledge, and the theoretical core may be regarded as quite limited when viewed critically; the rest is approximation with a different status of truth. Thus, even the "laws of nature" reflect an ideal state, and scientists function for the most part in a world where their investigations cannot fulfill criteria in which the special conditions of law are applicable, and their descriptions consequently fall into lower forms of knowledge (see, e.g., Cartwright 1983, 1999; Giere 1999).

The third conclusion is that we have witnessed a most interesting episode in intellectual history in which two orientations developed to a common problem. The first we might label as the "passing trains" modality, where critics and their subjects, while aware of each other, exhibited little interaction. From the scientist's perspective, one cannot reasonably argue against the progressive features of scientific investigations. Advancement over earlier theories is self-evident by elimination of error, improved predictability, the power of assimilating divergent matters of fact, and the ability to manipulate natural phenomena. What could one dispute about the "truth" and "reality" of a designer drug (e.g., allopurinol) that specifically inhibits a receptor or an enzyme (i.e., xanthine oxidase), which, in turn, initiates an entire cascade of physiological events that end in the reversal of a pathology (i.e., gout)? This is science as a successful empirical, pragmatic enterprise that *works*. That sociologists have shown that scientific method *essentially* (i.e., formally) does not exist hardly addresses the scientists' concern. They do what they can. The question of how is someone else's problem.

From this parochial perspective, sociologists, historians, and philosophers of science—those I have referred to as "science critics"—are committed to their own professional ends as *sociologists, historians, and philosophers.* Indeed, a strategy emerged to ward off the attacks of science defenders by asserting the professional interests of the critics of science as essentially their own concern. Accordingly, only a radical fringe sought to alter scientific practice or to insinuate themselves into the laboratory to become part of the scientific team. But the dominant stance was that of passing trains, so that the scientific community appreciated that the critical barbs being thrown by the critics would glance off their professional shields, and work would proceed without hesitation of introspection or self-critical doubt. Science remained committed to its own agenda as practitioners so defined it, and the critics, in any and all guises, could do what they wanted among themselves.

Finally, we might consider the deepest metaphysical challenges hidden beneath the polemics: If recent science studies critics effectively dismantled the epistemological project presented by staunch defenders of science, then what is the status of the metaphysical picture science has presented? If science's own value structure shows itself weakened by conceits of objectivity and neutrality, if its truth claims are viewed with a deeper skepticism and its theories threatened with dismissal, if its promises of factual knowledge prove insecure, *reality* collapses . . . at least the real as construed by what Western society has regarded as the paragon of reason. We never truly expected certainty, but we did expect progress and a reasonably reliable image of reality. What could take its place? When so viewed, the Science Wars take on a new urgency. The anarchists are at the gates, and what will be left if their attacks cannot be totally repulsed? If they breach the walls of the laboratory, then, truly, where are we?

As we have discussed, the "radical" critique emphasizes the dynamic character of scientific advances, where no proscribed method exists, and theories rise and fall not solely on their claims to objective truths, but are also beholden to hidden aesthetic and political forces. Revisionist descriptions of scientific practice have shown how the social institution of science has had a powerful effect on what scientists study and how they examine their objects of investigation. They have also revealed how the politics of science, both at the local laboratory and at the global arena, follow sociologically determined ways of knowledge formation similar to those found in nonscientific pursuits. Perhaps most importantly, despite all these caveats to the positivist picture of science, truth continues to reign, although some truths are more true than others; relativism holds no sway, and objectivity, despite the impossibility of its demands, remains standing in Western societies. *The* question then, given the limits of scientific rationality, objectivity, and the standing of truth, is: how does science function so effectively in its quest for the real? This last question, this final, impossible, irreverent query, continues to wag its weary head at us, for underlying the critical appraisals of scientific discourse reside profound philosophical imbroglios about how the scientific mind addresses the most profound of mysteries, what "mind" is and how it can know the world. So, beyond the epistemological quandaries and the intellectual realignments reflected by the Science Wars, profound metaphysical issues remain largely undeclared and dormant beneath the obvious issues at stake.

Conclusion

Ironically, opening science to a newly invigorated criticism collapsed the two culture mentality as a problem of scientific literacy, but at the same time, the criticism signaled the continued separation of scientists and humanists, not by lack of common language, but by different professional interests and, more importantly, philosophical orientations. (For instance, philosophy of science seemed to offer little, if anything, to practicing scientists, but the philosophical insights garnered over the past fifty years have had a profound effect on philosophy's own epistemological agenda. The same is true for the historians and sociologists in their respective fields.)

But did the separation of the contending parties actually result in no new synthetic position? A more circumspect view, one I will call the "open door policy," represents the alternative to the "passing trains" sociology, and from this position, another conclusion beckons. Historical and sociological studies have demonstrated beyond reasonable doubt that the working *practices* of scientific disciplines are both incompletely and inaccurately portrayed by the methodologies to which scientists officially subscribe. The challenges to the pretensions of the sciences that such discrepancies pose are surely not to be dismissed as illegitimate interventions from "outside" science. Rather, they arise from within the sciences, though from aspects ignored by most practitioners (Jardine 2000, 233).[10]

However, these epistemological issues should not obscure the deeper moral challenge, one that draws from science's own origins and the undeniable strength of its method. Considering the wide-spread skepticism about the character of rationality, the nature of truth, and the notion of objectivity, which together have formed a powerful attack on Reason, science offers its unique alternative as a paragon of knowledge and judgment. At the interface between a philosophy that confidently proceeds to discover the order of nature and a cultural ethos increasingly doubtful about universal truth, reason or objective knowledge, a rift opens. Part of the intensity of the Science Wars reflected the deep disunities of how reason is conceptualized. We saw this in the intelligent design controversy, and it appears in many guises of social policy, as we will consider in the next chapter. In a society that protects pluralistic values and disparate conceptual frameworks of understanding the world and the place of humans in it, informed citizens hold the fate of liberal democracies. These differences cannot be erased nor even resolved. Only liberal tolerance allows such diversity to coexist, and science perhaps makes its critical moral contribution by exhibiting standards of discourse, which are open and theoretically transparent, as a model of liberal dialogue. Self-awareness of this moral posture makes the scientist's ethical role more important as the unique qualities of her vocation become increasingly distinctive. Such moral self-consciousness places her within an even larger intellectual undertaking than the description of the natural world or its mastery. That humanist undertaking would again make a scientist a natural philosopher. For some, this calling would address a wider political and social agenda, and, as the next chapter discusses, when science becomes a participant in social policy, enlightened experts serve a critical role in democratic deliberation. Science as politics is the next step in the trajectory charted here: individual knower to community of researchers, and now to the social world to which they contribute.

5

Science in Its Socio-political Contexts

We can comprehend only a world that we ourselves have made.

Friedrich Nietzsche, *The Will to Power*

Scientific knowledge is neither value-free nor as "neutral" as Weber had hoped (see chapter 2), for the choice of investigation, the resources allocated to the endeavor, and the uses to which the knowledge is put, represents a goal-directed venture. Saturated with social purpose, the application of scientific findings informs decision-making affecting social welfare, economic wealth, and military strength. (*Neutrality*, of course, is different from *objectivity*, a topic discussed in detail below.) I am referring to such issues as:

(1) Should more prisons be built? Part of the answer depends on the role of rehabilitation, which, in turn, rests on assessing the degree to which genetic inheritance determines human behavior and then establishing the extent to which training or medication might alter such genetic determinism. Based on these judgments, behavioral sciences play a crucial role in setting the framework for establishing goals of incarceration.

(2) Given the costs of energy, what are the most efficient, safe, and environmentally conservative alternatives to fossil fuels? Part of the answer depends on basic research, part on technologies, and part

on economics. The science underlies each domain, not only in the development of applications, but also in assessing future effects.

(3) Concerning abortion, at what point in development does a fetus become viable, and what medical interventions are required to sustain the life of a premature infant? Clinical science frames the debate, which obviously is heavily laden with moral and religious influences, but the rationale for any position builds on a conception of life provided by medicine.

Application of scientific claims requires judicious and measured balance of what constitutes objective evidence and what information becomes enlisted in the promotion of economic, political, or religious power interests. In other words, policy disguised as scientific debate is often ideology by another name.

This point was amply illustrated by the George W. Bush administration, whose ideological positions insinuated themselves into diverse scientific arenas, including the termination of federal support for stem cell research and the denial of global warming (Mooney 2005). In the first instance, religious doctrine interposed itself to slow a dramatic scientific advance, and in the other instance, the myopia of business interests distorted the interpretation of earth scientists' warnings about significant climate changes. These two recent examples highlight how science becomes subject to the political and social currents that may buffet the walls of the laboratory. And in an eerie Orwellian sense, the attacks are made in the name of "sound science," a phrase coined in 1993 by a nonprofit front group for the tobacco industry called The Advancement of Sound Science Coalition (TASSC), which was formed to oppose the regulation of cigarettes. The term has been repeatedly employed as part of a lexicon for putting a pro-science veneer on policies that the general scientific community disavows. This rhetoric has allowed for the articulation and application of "peer review" criteria designed to obstruct policies that were at odds with Bush's political agenda (Mooney 2004).

Gary Trudeau, through one of his Doonesbury cartoons, eloquently captured our political temper in his typical trenchant fashion (2006). The cartoon pictures a diligent student working on a calculation and exclaiming in frustration, "Fudge! That can't be right!" After throwing away the calculator, he mumbles, "Drat! These pesky scientific facts won't lineup behind my beliefs!" Dr. Nathan Null, the White House "situational science" advisor, then enters. Null confirms the boy's insight and then opines:

> Situational science is about respecting **both** sides of a scientific argument, not just the one supported by facts! That's why I **always** teach the controversy like the evolution controversy, or the global warming controversy . . . not to mention the tobacco controversy, the coal slurry controversy, the dioxin controversy, the everglades controversy, and the acid rain controversy.
>
> The earnest student replies, "You're right, Situational Scienceman—I'll never trust science again! It's just too controversial!" The cartoon ends with the advisor peering out at the reader, "Stewie gets it now folks! Do **you**?" (emphasis in original).

The politics of science has captured center stage in many policy debates, and in the most recent attempts to curtail the use of science, we witness a political shift in these struggles: the stage has been taken from those on the left, who a generation ago opposed nuclear power and promoted environmental protection, by a newly inspired conservative movement that would block stem cell research and genetic engineering for their own ideological reasons.

Our consideration of these matters begins with two observations. First, the "package deal" of *doing* science and *placing* science within its intellectual and social contexts argues that science and its study as a human activity cannot be separated. This interdisciplinary effort arises because neither the laboratory nor technical discourse can circumscribe the boundaries of science (Gieryn 1995). The findings seep into applications, which affect our material culture, medicine, the military, and virtually all aspects of Western societies. Only an educated public can make appropriate use of the fruits of scientific labor; thus scientists and policy makers must closely coordinate their efforts to reap the greatest harvest from the investment made in research.

Second, the critique of science is essential to its flourishing. Science gains its place at the table precisely because of its power to define a competing worldview. The "naturalization" of humans, from the evolution of species to the biological character of the mental, testifies to how successfully scientific explanations have been translated into potent theories of human nature and society. However, notwithstanding the effective penetration of scientific theory into notions about the nature of our social and psychological existence, a careful scrutiny is required to apply the conceptual lessons appropriately. Closely linked to such applications, the converse operation is also necessary, namely, a critical view of the truth claims made by scientists. With these critiques, philosophy and history of science find their most pressing calling. As already discussed, on this broad view, science is part of a larger historical development of humanism and ultimately finds itself in its service to pursue means of achieving human goals and ideals.

Boundaries of inquiry define the realities that science explores and describes, and those boundaries, which we have already considered in their intellectual context, are here explored as a political issue. If the period bracketed by the Second World War and Kuhn's *Structure* is labeled "Two Cultures," then our own era has been "One Culture," and radically so. We now appreciate how firmly science is embedded in its social matrix. To gain a perspective on this issue, we must better appreciate how the "network" of scientific activities, findings, and ethos have deeply penetrated culture and in turn have been molded by that culture. As Bruno Latour describes the phenomenon, science is "blended," by which he means to emphasize how artificial it is to attempt to dissect science away from its supporting society as some circumscribed intellectual and technical activity (Latour 1993, 1999). Beyond the social analyses that divide complex cultural activities in order to examine them as entities, contemporary social theory avers again and again the fundamental inter-contextualization of complex Western social institutions in which science takes part. For our purposes, the network is a complex of values that both plainly and covertly guide science's reason.

The so-called model of "Extended Translation" (Callon 1995) displays science as a vastly intricate network operating in a communication system that connects the various levels of its discourse constitutively, as well as outwards through a supporting social milieu. By and large, science is erroneously (or perhaps just superficially) regarded as having its own sophisticated domain, and ventures forth from its insularity with caution against a potentially intruding public. The lobby promoting scientific research and training routinely expounds the obvious benefits of scientific progress for technology and medicine and often treats the layperson to a dramatic presentation of Science at Work. Whether through pictures of a comet crashing into Jupiter, the pinpoint accuracy of smart missiles, or the *in-vitro* fertilization of a sterile sixty-two year old woman, Bacon's espousal of science's promise to better society is constantly reiterated. Not solely the business of scientists, their findings and disputes become intertwined into the interstices of seemingly disinterested parties to become integrated within society as a whole.[1]

The political uses of science appear throughout society. I construe "political" as extending from social policy to economic resources and allowances; from definitions of normal and abnormal to prescriptions of health and disease; from education and definitions of knowledge to describing the limits of knowledge and the place for religious belief; from depictions of social organization to the psychology of cognition, emotions, and motivations. In short, scientific explanations are inseparable from definitions of who we are, prescribing what we do and explaining why we do what we do. Science thus moves well beyond the laboratory to the newsroom, the legislative committees, the classroom, and the poll booth, with scientific advisors comprising the "fifth branch" of government, serving both domestic policy makers (Greenberg 1967; Gieryn and Figert 1990; Jasanoff 1998) and those designing foreign policy (Doel 1997). Indeed, boundaries have been difficult to determine for those who would circumscribe science's activities in order to characterize its influence, costs, and social contributions. The most judicious conclude that boundaries have vanished, if they had, in fact, ever existed.

Obviously epistemological limits are always at play in the boundary debate. The first boundary issue pertains to determining whether social phenomena are amenable to scientific scrutiny and criteria. In other words, what are the relations between the natural sciences and the human sciences? What aspirations should the human sciences seek as science? Or, more fundamentally, are the social sciences science in the same sense we think of physics, chemistry, or biology? The second boundary question pertains to the social contextualization of the natural sciences: how do we understand science sociologically?

To approach these questions, we return to the critical provisos we have already outlined about the professed ideals of scientific inquiry. These include the ambiguities regarding objectivity, the delineation of the limits of natural scientific theory, the antirealist concern that epistemic judgment depends upon local contexts of use, and the pragmatic attitude that dominates scientific inquiry. Previously, we explored the theoretical aspects of this problem, and here we turn to the "boundary question" as viewed by a classic sociological perspective, namely, how the *institution* of science might be characterized. The struggles of the Science Wars extended to this political domain, not only as all sides felt the repercussions of highly iconoclastic attacks on the foundations of knowledge, but also as the debates extended to the more mundane issues of how much money governments should budget for basic research. From this vantage, science studies critics are engaged in a broad political process.

Recent debates notwithstanding, intense discussion about the moral standing of science assumed its distinctive American voice with John Dewey, who advocated in the 1920s the integration of science with the larger moral-political domain as part of his liberal agenda:

> When all is said and done in criticism of present social deficiencies, one may well wonder whether the root difficulty does not lie in the separation of natural and moral science. When physics, chemistry, biology, medicine, contribute to the detection of concrete human woes and to the development of plans for remedying them and relieving the human estate, they become moral; they become part of the apparatus of moral inquiry. . . . (Dewey 1920/1948, 173)

Such a melding would directly benefit moral-political discourse and provide the natural sciences with a better articulated sense of purpose:

> Natural science loses its divorce from humanity; it becomes itself humanistic in quality. It is something to be pursued not in a technical and specialized way for what is called truth for its own sake, but with the sense of its social bearing, its intellectual indispensableness. It is technical only in the sense that it provides the technique for social and moral engineering. (173)

As we will discuss in the Conclusion, Dewey identified this integrative project as the key philosophical problem of his age (and I believe it remains ours as well), namely to bring the fruits of scientific investigation to their full human potential and the transfiguration of moral discourse from thin pedantic argument to intelligently informed debate:

> When the consciousness of science is fully impregnated with the consciousness of human value, the greatest dualism which now weighs humanity down, the split between the material, the mechanical, the scientific and the moral ideal will be destroyed. (173)

Following Dewey, later commentators like James Conant (1953) and J. Bronowski (1956) sought to situate science firmly within a tradition of public discourse and humane values, but others, like Snow, were preoccupied with the disjunction of intellectual life and the cultural segregation of science (1959/1964). The growing autonomy of science, residing in its technicality and esoterica, seemed to go unchecked and to be unchallengeable. Snow's argument resonated strongly with those suspicious of science's cultural power, and more particularly, the threat of nuclear madness. Although Snow was arguing a different agenda, the contrast between then and now is striking, for he wrote at a time when science was generally regarded as some distant colony, aloof, if not markedly separate, from the rest of society. But now, most commentators see science as fully incorporated into its larger culture to the point that it is fighting to regain its own "sense of self." The pendulum has swung wildly in the past four decades: science cannot be given the status of some autonomous social activity, or even of a distant territory that enriches the mother country. Instead, science has become constitutive to our very selves, interpreting through its own refractions issues heretofore left to ethics, religion, and philosophy.

Disentangling Facts and Values, Again

The boundaries of science may be drawn in various dimensions, and now we will consider contemporary views of how scientific knowledge is applied in public policy decision making. As already discussed, scientific and moral discourses are historically interwoven (Shapin 1994), and despite attempts at maintaining science's neutrality, the lines segregating science from politics are characteristically blurred. So, while science has typically been cast as an actor in political dramas, as if the science dictates one set of choices versus another, the actual script is hardly so well defined. More often than not, the scientific finding is a matter of dispute and the "facts" then serve as instruments for advocates of one position or another. At this juncture we witness how the constructivist critique makes its bite: natural realities are configured by various extra-curricular factors and thus objectivity becomes contested. Simply, as facts are translated into the domain of policy, education, self-understanding, and so on, they assume disputed meaning and the objectivity that discovered and reported them as knowledge becomes another variable in the political contest.

Despite these ambiguities, two basic views of the scientist dominate the public's image of the investigator. The first we will label as "above the fray." This persona presents herself as an inquirer for Truth and Reality. She resides within the laboratory, shielded from the messy debate of how the fruits of her labor are applied (whether in warfare, medicine, or technology at large) or how her findings might precipitate dire consequences for the environment, society, or the individual. Insulated in their work from daily tribulations, "above the fray" scientists explore nature's secrets, oblivious to the political, social, and economic needs of the supporting culture. Accordingly, the notion of "forbidden knowledge" (Shattuck 1996) contradicts the very ethos of the scientific enterprise; research must proceed in whatever direction curiosity and need dictate.

The alternate scientific ego, "Faust," deplores the conceits and moral irresponsibility of such a neutral stance. Hostile critics decry the scientist's false neutrality that feeds a ravenous techno-industrial society built upon a dispassionate rationality, which is employed for knowledge divorced from *humane* values:

> Bacon went in search of a philosophy of alienation. They [scientists] broke faith with their environment by establishing between it and themselves the alienative dichotomy called "objectivity." By that means they sought to increase their power, with nothing—no sensitivity to others or the environment—to bar their access to "the delicate mysteries of man and nature." The cult of objectivity has led scientists and the general public, to think of everything around us—people and biosphere—as "mere things on which we exercise power." Objectivity is in practice a cloak for callousness. (Roszak 1972, 169)

This vision portrays the scientist's political overtures as no more than an undisguised effort to promote the scientist's self-interest.

So two personifications vie for the public's understanding. On the benign view, the "common" people reap the material harvest of science's efforts. The cynical appraisal argues that Western societies have made a Faustian bargain, the deleterious effects of which cannot be controlled. In the first case, while the scientist remains divorced from the social-political world, the product of her labor nevertheless percolates through its technological application and policy debate to benefit society. Objectivity then, assumes a particular social value. At best, such an agent is regarded as benign and trustworthy. The evil alternate also claims acknowledgment; the scientist emerges out of the laboratory in political attire to engage other contestants in public debate on social, economic, edicational, and cultural matters. In that case, she not infrequently appears as a trespasser who would promote her own fiendish vision upon her fellow citizens. From Frankenstein to Dr. Strangelove, this image has captured the public's worst fantasies—a personification of a disparity between how the nonscientist sees the scientist and how the scientist sees herself. This depiction is disturbing, to say the least.

Such diabolical caricatures are, by and large, straw figures constructed by various lobbies for their own purposes. Once the polarization of scientist/citizen is abandoned, everyone, scientists and laypersons alike, turns out to be aligned on the political spectrum, advocating various positions with varying commitments to scientific arguments for support. In these debates, the uses of science cannot be neutral. When the findings of science are used in social debate, those findings must be interpreted and extrapolated to decisions about public policy. Science's *neutrality* is lost because human assessments are imposed onto what, left isolated, might otherwise claim no value. The *objectivity* of science depends on regarding nature as holding no value. Values are rooted in human needs and desires, whereas nature, stripped of qualities, teleology, and meaning, is left secularized, value-neutral, disenchanted. (In the next chapter, we explore this assertion.)

The crucial philosophical distinction upon which the discussion rests concerns how to differentiate "what ought to be" from "what is" (Hume's Law).[2] The attempt to free fact from value was to liberate science from its medieval theological roots, and remains the lynchpin for scientists pleading autonomy under the rubric of objectivity, as well as for their critics, who decry the violation of neutrality of science, which obviously serves particular social agendas. But as Robert Proctor has cogently observed:

> neutrality and objectivity are not the same thing. Neutrality refers to whether science takes a stand; objectivity, to whether science merits claims to reliability. The two need not have anything to do with each other. Certain sciences may be completely "objective"—that is, valid—and yet designed to serve certain political interests. Geologists know more about oil-bearing shales than about many other rocks, but the knowledge is thereby no less reliable. Counterinsurgency theorists know how to manipulate populations in revolt, but the fact that their knowledge is goal-directed does not mean it doesn't work.
>
> The appropriate critique of these sciences is not that they are not "objective" but that they are partial, or narrow, or directed towards ends which one opposes. In general, knowledge is no less objective (that is true, or reliable) being in the service of interests. (1991, 10)

This critical stance is based on the assertion that science as practiced is not a free-standing enterprise, but is firmly based in the social and subject to the needs and values of its supporting culture. This public domain of scientific research refers not only to the renewal and support that society gives scientific institutions, but to the recognition that science serves diverse interests in a political culture striving to balance competing economic and social forces. From this perspective, the relevant issue, beyond defining the social origins of knowledge, is the requirement for a philosophy that focuses on the forms of power in and around the sciences: "Why do we know what we know and why don't we know what we don't know? What *should* we know and what shouldn't we know? How might we know differently?" (Proctor 1991, 13). In short, citizens must be cognizant of the complexities of science as politics and not confuse the interpretative applications of science with the business of scientific inquiry. So while it is appropriate that extrapolations be made—after all, science is the paragon of knowledge acquisition—interpretations must be *acknowledged as interpretations*. We might well assume a cautious attitude towards scientific neutrality, because scientists, perhaps unknowingly, may hide behind an objective mask to pursue unstated ideological goals. This issue frames much of the radical constructivist critique and is the one point that arouses the most hostility among defenders of science, who would argue that in these perverse uses of science, we witness the political usurpation of what should be an autonomous endeavor. In characterizing and acknowledging scientific accomplishments, we must still realize limitations, false applications, inappropriate expectations, and the dangers of unbridled scientism,[3] where the limits of science, the contingencies of its methods, and the boundaries of its applications are not understood.

The debate over science's place in society has a long history. Four hundred years ago, Bacon astutely recognized that scientific knowledge confers social power and technical mastery, as science turned from "a contemplation of eternal truth to science as instrument of social progress" (Rorty 1991a, 33). Indeed, some would argue that science is only that which can be traced to some technological advance. As purveyors of technology and power, scientists may be regarded as an instrument of political authority, whether in monarchial, totalitarian or democratic societies. Further, as the scientific community has grown in the past sixty years, scientists and their supporting industry have increasingly been characterized as an interest group advocating their method and product for their own economic purposes. Unfortunately, there are dramatic historical examples of perverse political manipulation of science in the case of Stalinistic genetics (the Lysenko affair) and Nazi racial views, but these are regarded by radical critics as only more obvious examples of the political nature of even normal science.

Argument over the use of science for political programs or the development of social policies has had several lives in the twentieth century. During the first four decades of that century, debate enlisted the latest scientific findings to argue about the equality of women, the merits of social Darwinism, the logic of eugenics, and the rationale of racial hygiene. Following Hiroshima, the relationship of science and political ends took on new moral urgency, which revolved around Ban the Bomb movements through the 1950s. During the 1960s, political activism that was organized to protect the environment gained popular support, and in the 1970s, fuel efficiencies, pollution, and genetic engineering each commanded broad public discussion about the application of scientific knowledge to achieve rational social policy. Debates focused on how political ideology might color not only scientific goals, but the very research programs promoted as value neutral. While this question was bolstered by renewed criticism from European post-World War II social philosophy and American feminism, the origins of the controversy about science's neutrality are more directly traced to the birth of German sociology a century ago (already discussed in chapter 2). Today our current fears about the consequences of global warming bring this message to the forefront of public discourse. As we struggle to identify the consequences of past practices, we turn to scientists to predict the future of the environment and to provide options that will alter our everyday existence. To state the obvious: science pervades everything. We do well to understand its workings better.

Science as Politics

During the 1930s, sociologist Robert Merton elaborated four principles, or norms, by which science might be evaluated, a theme brought up to date by Sheldon Krimsky (2003):

(1) *universalism* refers to how science transcends national, cultural, or institutional boundaries and thereby is a shared activity of humanity at large;
(2) *communalism* respects public ownership of the products of scientific investigation, especially in regard to the extensive government support provided to research;
(3) *disinterestedness* requires scientists to perform and to interpret their work without considerations of personal gain; and

(4) *organized skepticism* demands that scientists suspend judgment of findings and their application until deliberate review is made.

Closely tied to neutrality, these last two principles suggest that scientists should leave their ideological hats and political jackets at home.

Controversies surrounding public policy concerning investment in major scientific projects that are touted as the penultimate, if not the ultimate, peak of scientific development (e.g., for molecular biology by Gilbert 1992, and for particle physics by Weinberg 1992) are largely connected to three recent assessments. First, the similar programs of the past, such as Nixon's "War on Cancer" and other overly optimistic projects that promised to deliver unrealistic solutions, have had disappointing results. The political response has been a growing concern that resources should be more carefully allocated toward directed application and more modestly achievable goals. Second, science and its uses are not easily separated, so policy makers have become increasingly alert to the possible genesis of new industries and the growth of those already established. Imposing potential limits on the growth of science has led some critics to propose that indeed there are forms of "forbidden knowledge" (Shattuck 1996). The most obvious current case in point concerns genetic engineering. And finally, perhaps the catchall of a new prudence, the widespread recognition that the positivist ideals of the scientific method have been weakened means that the progress and application of science must be viewed with more circumspection (Kitcher 1993). In other words, scientific truth claims are subject to a more skeptical assessment.

Investigators easily move from the laboratory into society-at-large by advising and directing public policy (e.g., Ashford and Gregory 1986; Jasanoff 1998; Greenberg 2003). Here we witness the makings of a political philosophy of science, for in the judicial or regulative advisory position, the scientist plays to both scientific and broader humane concerns. The expert must offer her best professional opinion regarding risk but at the same time recognize that any ostensibly objective judgment may be biased. The first caveat is to acknowledge that interpretations of scientific facts are not necessarily impartial, and that scientists carry complex personae. Prejudicial judgment has many sources, some intrinsic to the science per se (e.g., a commitment to a particular theory that may be in dispute, or confidence in a certain methodology that may be questioned) and others extrinsic to the narrow confines of the laboratory (e.g., a political affiliation or religious belief that might seek scientific support).

Dispute, whether about the safety of a drug or the environmental impact of a chemical, generally reflects a continuum of scientific certainty along which facts may assume various meanings. This point reiterates Proctor's concerns, which ultimately arise from science's own quest for an elusive certainty. Accordingly, the reasonable doubts raised by philosophers, historians, and sociologists concerning the epistemological foundations of scientific inquiry have social consequences, namely an assessment that appreciates the complex factors that lead to any scientific conclusion.

The scientist typically is criticized not for her scientific claims but, more often than not, for her political stance. She will increasingly be placed in the contentious role of social activist under the cloak of her professional credentials, as scientific questions are posed and the answers formulated become central to the public debate, for instance, about human nature or environmental policy. Of course, such controversy will also expose the vulnerability of scientific knowledge, that is, its provisional and tentative character. And those who view science as a normative activity may find it painful to see investigators dragged into the trenches of current political warfare, where bloody contests ensue as one group of scientific experts is pitted against another. Scientists debate among themselves about data and theory as a matter of their normal discourse. The public forum, often exposes these same scientists as less certain (and thus less authoritative) than the lay public might wish.

Science does have legitimate claims to rationality and objectivity in pursuit of its narrowly articulated objectives, and these objectives hold important promise for human welfare. But when scientists engage in public debate on social questions, their authority is subject to different rules of inquiry. On virtually any controversial social question, from abortion to waste control, scientific testimony typically aligns itself on both sides of the issue. Argument is usually pitched between opposing experts, and citizens watch the spectacle of a contest concerning whose data are more valid or on what basis such data might be enlisted. Skepticism about scientific certainty, or at the very least about bona fide knowledge, opens the door to decisions determined by criteria other than what are normally construed as "scientific." These decisions may, of course, be determined by moral, legal, or, frankly, political rationales, and the scientist then becomes a bit player in a larger social drama.

While guidelines for policy makers may offer a practical solution, albeit imperfect and limited, activists such as Feyerabend (1978) argue for more radical citizen participation in judging science, whether in regard to its technological products or its testimony, which may have broad social consequences. As science continues to amass further sophistication and complexity, and its products continue to dramatically change daily life, citizens will be increasingly concerned that science is too significant to be left to experts alone. Indeed, government oversight and citizen participation in policy decision making has progressed steadily from the jocular acceptance of the authority of scientific experts ("There is a current saying among government supporters of research that scientific research is the only pork barrel for which the pigs determine who gets the pork" [Kenneth M. Watson, quoted by Greenberg 1967, 151]) to vigilant scrutiny of research budgets and the purported benefits of the programs funded. The growing presence of governments in regulating the laboratory, protecting human subjects, examining research budgets, and monitoring investigators reflects the various demands to control the course of research (Greenberg 2007).

The growth of the environmental movement forcefully articulates broadly held sentiments that science must not be deified into an unassailable ideology of the state or corporate interests. (For example, Dow Chemical's slogan, "Better living through chemistry" would be aggressively challenged from this point of view.) Not only would such critics regard science with a skeptical eye, they would further argue that other forms of knowledge are equally important and offer legitimate bases to assess scientific advance. Some regard such arguments as, at least incipiently, "antiscience." But in seeking to examine the underlying claims made by science for formulating our construction of reality, the debate must strengthen the scientific project itself. The issue is not scientific validity; after all, none could reasonably deny scientific achievements. But Feyerabend and like-minded critics endeavor to place those accomplishments within their proper truth claims and to reassert a more encompassing philosophy. And when one also considers how scientific method might serve broader social values, the examination demands a proper assessment of the scientist's role. Scientists are expert consultants, not infallible authorities. We must use their knowledge critically as part of complex social deliberations (for examples of those who do, see Collins 1992 and Jasanoff 1998).

In sum, science functions as a political force in which scientists vie with other interest groups to garner public support. Detractors have attempted to depict the proponents of such projects as the super collider as self-aggrandizing competitors contending for scarce economic resources within a political arena. Science then becomes another project for debate, just like subsidies for milk, pork barrel patronage for public works, or special tax breaks for struggling industries. In this context, scientists occupy no sacrosanct position and must pit their lobbyists against those of other interest groups similarly seeking government subsidy. The same rules apply, and the same utilitarian factors determine the outcome. On these playing fields, science is just another participant in contemporary power politics.

Science and Human Nature

The kinship of humans to other animals has preoccupied the human sciences since Darwin, as biologists have sought to inform and influence the rational understanding of human nature in the context of our evolutionary origins. For instance, evolutionary explanations are now offered to elucidate complex social behavior like altruism or the economics of commerce (de Waal 1996; Ridley 1997; Segerstrale 2000a). Recent hot topics include: Is IQ racially correlated and therefore based on inherent biological differences? Does homosexuality arise as a result of a gene or a cluster of genes, and is it therefore biologically determined? To what degree is schizophrenia, or any other mental illness, genetically determined, and by extension, is behavior or personality neuronally hard-wired—and thus genetic? Beyond such catholic questions, biological criteria are sought for what is "beautiful" (Rentschler et al. 1988); E. O. Wilson, for example, postulates that some innate (programmed) belief in the divine grounds ("explains") religious practice (Wilson 1978). Indeed, the very way we think and the cognitive basis of language are now heatedly debated as biological phenomena, best understood in the context of evolutionary pressures (Deacon 1997; Pinker 1997).

These issues do not simply reside within biology but have surreptitiously crossed over into the domain of human values where moral orientations and human prejudice heavily influence how these matters are framed and discussed. This transition from ostensible objective scientific evaluation to social opinion has been made effortlessly in current polemics about the environment, education, race, abortion, sexuality, mental illness, criminal deviance, or any number of other social questions that have sought biological, especially genetic (Lewontin 1991), models of explanation. At the most basic level, the very possibility that science might answer such questions already frames the kinds of responses that will be offered.

Undoubtedly, science is a powerful ally that many parties invoked to support or challenge contending positions in social policy debates involving the life sciences. If indeed we seek rationality in adjudicating complicated social questions, it is expected that we would call on the most sophisticated and informed scientific opinion to derive the best solution. But debates are ongoing about how, or even whether, biology should be used to formulate the human sciences. To what degree biology defines human nature—for example, the degree of genetic determinism that might program complex human behavior—has a profound influence on ethics. In effect, the question is to what extent people are responsible for their behavior if biology dictates that they are little more than their genes, as opposed to autonomous, free moral agents. In fact, we might well question the status of a psychology that is reduced to aberrant biochemistry. How do we judge and punish the criminally insane? What is addiction? What are the educational "rights" of dyslexics? Is homosexuality "normal"? What, indeed, is normal? An answer to the last, and most basic question remains elusive, not only because of the complexity of human behaviors, but also because the "normative" is always contended (Canguilhem 1989; Foucault 1980, 1994).

It is impossible to totally separate current scientific appraisals from the rational construction of social choices and mandates, which, in turn, reflect deep cultural bias and tradition. The epistemological and moral debates are intertwined, supporting each other in obvious and not so obvious ways, so that the boundaries which would separate them endlessly shift, if not disappear. Scientists have become willing actors in this discussion. They often find themselves choosing a line of inquiry that posits an ideological endpoint with profound social ramifications. Thus, for example, the Human Genome Project (HGP) received financial and political support ostensibly to develop better technology for nucleotide sequencing and information processing to construct genetic maps (most immediately) and to serve as the foundation for advances in basic molecular biology. But it also has been enlisted to identify genes for various "social diseases" such as alcoholism or violent behavior. Thus the HGP has been trumpeted as both a Holy Grail for molecular biology and the Rosetta Stone for a future biology of human nature. Supported by many camps, not all of whom share the same agenda, the HGP is both science and politics.

The promise of genetic medicine is based upon deterministic causation. Simply, for its advocates, "the gene" initiates a string of behaviors that might be altered if we understood the initiating event (the aberrant gene). Accordingly, behavior, no matter how complex, becomes a problem for genetics. Moral responsibility then assumes new facets and the ripple effects throughout society apparently have no end.

Perhaps the basic challenge concerning the future of science as a social institution pertains to where the boundary between the laboratory and its surrounding culture is drawn. As the Science Wars so well demonstrated, this question has profound political overtones. The radical constructivist critique arouses the most hostility among defenders of science because they understand that to deconstruct scientific objectivity undermines the authority of science to frame public opinion. Such critical depictions of science are, according to defenders of an insular science, patently perverse and represent an ideological usurpation of what should (and can) be an autonomous, if not value-neutral, endeavor. Some constructivist critics counter that, as often as not, a particular research strategy has, within its very foundations, a bias (although usually left unstated or unrecognized by the scientist) that has broad social ramifications beyond its narrow research agenda. A case in point is the support genetic reductionism gives to a particular kind of biological determinism (Tauber and Sarkar 1992). So when critics railed against the HGP, they did so not only because of misgivings about its scientific strategy, but also because they perceived that it consisted of more than the stated direct purpose of mapping and sequencing genes.

Molecular biology has an undeniable authority when it remains within a narrow domain of inquiry, but extending the results of the HGP to social policy provides a vivid illustration of science's profound influence on culture. The nagging question endures regarding the extent to which science should enjoy such authority when its findings are projected or applied in domains so widely outside its more orthodox purview. This is the same question Weber posed, and it remains as volatile today as a century ago, and for the same reasons. The detractors' general concern about genetic reductionism draws from their resistance to the underlying genetic determinism of this scientific program, which in turn leads to a partiularly noxious, deterministic orientation toward human nature (Wilson 1978, 1998; Lewontin 1991; Tauber and Sarkar 1993). This is an ideological argument, resting on a complex array of philosophical orientations. To advocate resolution of contingent and complicated interactions of environment and heredity—when such a determination is impossible—reveals only opinion, not scientific knowledge. The stakes are high, for the vision adopted seeks to determine the way we regard ourselves both individually and collectively.

In sum, the polemics that swirled around the HGP when first proposed—those ethical and philosophical issues that eclipse the technical questions of how to effectively map genes or process the enormous quantity of data—may be fairly regarded as an example of science's political persona. The HGP amply illustrates that science is not simply prescribed by laboratory-based activity seeking "stubborn facts," but also includes two clearly declared political activities. The project is both a lobby to accrue government support for molecular biology and genetics and a philosophical debate about the merits of genetic reductionism. The debate then becomes one about public social policy in each venue: political and financial support for a particular branch of science and the application of a particular philosophical orientation to social issues.

The relevant issue regarding science's politicization beyond defining the social origins of knowledge, is the requirement for a philosophy that focuses on the forms of power in and around the sciences. It is, after all, naive to regard investigators as somehow isolated from their larger society and confined to their immediate research concerns. After all, while guarding their own domain, scientists still seek to inform and influence the political agenda of social policy as a reasonable extension of scientific knowledge. In the arena of social policy, the epistemological and the moral domains are inseparable because we integrate them as informed opinion on a complex continuum between our search for "what is" and our aspirations for "what ought to be." Again, firm boundaries fail because the fact/value distinction fails, which should make us pause and consider anew how a politics of science might emerge (see, for example Rouse 1987; Longino 1990; Gilbert 1997).

The independence of science is crucial to its health, so that any attempts to guide the direction of science by political or economic controls must be seen as part of a political process—insidious, necessary, and thus particularly difficult to analyze, forecast, and direct. Because of the surreptitious use of science for various harmful political agendas in the past century, liberal democracies have become particularly attuned to the dangers of a usurped objective authority. The character of science in this broadened view—no longer just a laboratory effort, but a complex social institution that impacts other cultural activities—remains a critical issue for the future.

Yet, despite cautionary provisos, we still seek authority, if not certainty, in our public debates. So in the very act of defining ourselves, the scientific view, with its strong claims to objectivity, is used to displace and outweigh other modes of discourse. The dialectic is of course bidirectional because our social and ethical ideologies may also color scientific interpretations of the nature of human psychology and social behavior. But of the two vectors, we more clearly appreciate the influence of science *on* culture, and as the authority of science has grown, its influence on human nature has increased in parallel, and from this position, we witness biological theory applied to the moral domain. For example, if homosexuality is regarded as biologically determined (a scientific judgment), and if biological determinism translates into psychological and social determinism (the conclusion of a human science), then how might we regard such behavior as deviant (a final moral determination)? To make homosexuality a transgression or an aberrancy, one must either use criteria other than science's (e.g., religious or ethical) or attempt to undermine and refute the science used to reach this unwanted conclusion. Increasingly, the course of employing other kinds of knowledge or rationalities becomes less tenable, and social debate is contested on scientific grounds where the objectivity/neutrality distinction must be carefully scrutinized. This complex dialectic of science affecting our moral stance and our moral views subtly directing science is, at its heart, the problem of placing science within its cultural context.

Political challenges are to be expected because the interrupted boundaries between science and other social activities are not readily defined. This problem is perhaps less epistemological than ethical, for how the borders are drawn is based on choice; choice is grounded on value, and value is a moral category informed by understanding. An educated public is the best assurance that science will be protected, promoted, and understood in its full complexity—for what it offers, as well as for what it cannot provide. Let us consider a case example below.

A Case Study: Environmentalism

Claims for a Science-based Ethics

How is nature, and especially our relation to it, meaningful to us, and in the context of environmentalism, what are our responsibilities for the use of nature? Our age has witnessed an increasing separation from the natural world. Both individually and as a culture, we spend a large part of our national fortune on studying natural phenomena for our economic and social welfare, and the technological product of that endeavor has had a tremendous price. A biocentric ethic has emerged in the attempt to promote a humane philosophy of nature.[4] Instead of fully exploring the moral philosophy that accompanies the environmental program, I seek here to show how values embedded in biology have led some to interpret nature with this environmental sensitivity. My argument explores a fundamental confusion concerning how that "green" position draws from biology, where the life sciences have been co-opted to serve an ideological agenda. Environmental ethics may be legitimatized by several routes, but I am less concerned with the moral logic involved than to demonstrate how facts and values interplay in the creation of the rights of nature movement. Ecological ethics is an important case example of how the fact/value distinction has been interpreted—defended or collapsed—in the political arena, and thus demonstrates how easily the line separating epistemology and ethics is crossed. In short, I wish to outline a "moral-epistemology" at work.

Much of what passes as ecological ethics claims that, unlike other ethical ventures based on religious or metaphysical foundations of belief, these ethics are rooted, in fact "proven" or "demonstrated," by the incontestable facts of a new science—ecology. Accordingly, if laboratories can demonstrate the deleterious effects of aerosols in the atmosphere or particularly toxic chemicals in our rivers or automobile fumes in our cities, then it is senseless, if not immoral, to continue along the path of environmental degradation. From this perspective, there is a seamless joint between the findings of ecologists as scientists and the values drawn from their studies.[5]

Why and how is the claim made, and is it legitimate? There are many reasons that might be conjured to support an environmental ethics, and I will only mention one and then delve into another. The first is a utilitarian imperative, which seeks to root ethical decisions in an objective accounting of gains and losses. By invoking the power of an objective science to attain some rational ideal, seemingly innocuous interpretations of the objective data then segue into a moral conclusion. But such interpretations, of course, have a moral setting, which while orietned by our scientific understanding, is hardly neutral. As Luc Ferry puts it, there is a syllogism at work:

(1) The biological sciences, including ecology, have disclosed that organic nature is systematically integrated:
(2a) That mankind is a non-privileged member of the organic continuum, and
(2b) Environmental abuse threatens human life, health, and ultimately happiness, therefore
(3) We ought not violate the integrity and stability of the environment (1995, 88).

Like any utilitarian argument, one may contest its relative merits. In this case, the ecological syllogism fails to acknowledge the relativity of the ideal, "health." Moreover, it asserts a nonnormative analysis of what we, de facto, are supposed to love or abhor. "The ethical criteria becomes identified with what empirical anthropology teaches us about human . . . psychology" (89). In assuming that we all share the same values concerning what is healthy, any deviance then becomes pathological. And perhaps most compelling, utilitarian arguments easily fall prey to the political contentions of competing interests, for example, President Bush's original dismissal of curbing petroleum emissions (in the face of global warming) in order to protect the U.S. economy.

Because of the difficulties encountered by the utilitarian position, a second argument is often made that more aptly illustrates the theme we are exploring, one that would assume the moral high ground. This position builds on the idea that life has a *telos*, that is, goals by which organisms organize their behavior and physiological organization to support end-seeking functions. The argument begins with asserting that organic life possesses certain objectives, for instance the preservation instinct, independent of our humane or subjective sentiments. From this position, teleology blossoms like a prolific bush, sprouting assertions in many directions. Let me sample a few representative opinions.

First, consider how an empathetic element is introduced. For instance, Paul Taylor writes how science documents the life cycle of the species, its ecological interactions, and so forth, and at the same time recognizes the uniqueness of each individual, which he believes may easily convert to a moral perspective. As he writes:

> This progressive development from objective, detached knowledge to the recognition of individuality, and from the recognition of individuality to a full awareness of an organism's standpoint, is a process of heightening our consciousness of what it means to be an individual living thing. We conceive of the organism as a teleological center of life, striving to preserve itself and realize its good in its own unique way. (Taylor 1986, 120–21)

The telos of the organism defines what is good for it, and the shared sense that organisms are individuals, like human agents, confers a moral standing to them. Simply, teleological centers of life serve as the foundation of value. And the entire enterprise rests on the science of biology, for as Taylor asserts:

> certainly our acquiring scientific knowledge about certain kinds of animals and plants can help us enormously in the attempt to understand objectively the everyday existence of particular individuals of those kinds. (126)

Thus a biocentric value is inserted into the biology, putatively an objective, fact-based science.

Note what is occurring: the purportedly objective science is in fact documenting *value* as constructed within an evolutionary and physiological context. In this vein of thought, Holmes Rolston writes that the organism is an "axiological system," an "evaluative system" as it grows, reproduces, repairs its wounds, and resists death. He then slips in the V word, value:

> Value is present in this achievement. Vital seems a better word for it than *biological.* We will want to recognize that we are not dealing simply with another individual defending its solitary life but with an individual having situated fitness in an ecosystem it inhabits. Still, we want to affirm here that the living individual . . . is per se an intrinsic value. . . . The organism has something it is conserving, something for which it is standing: its life. (1988, 100)

Then the ethics become explicit:

> There seems no reason why such own-standing normative organisms are not morally significant (1998, 100). . . . [Thus a] tree has a *telos* before the logger arrives, and the logger destroys it, it is *auto-telic*, it has a law (Greek: nomos) on its own [= *autonomos*]. (105)

Telos, essentially a descriptive mode that biologists employ to describe function, Rolston has extended to a moral category. What he calls the good of the organism, namely striving to preserve itself, a biologist, even employing a teleological orientation, would say is a survival behavior following the rules of biological fitness, where fitness refers to the ability of the organism to ensure the continued existence of its genes in future individuals. This is a precept of evolutionary theory, but Rolston assigns a moral value to it:

> Value is not just an economic psychological, social, and political word but also a biological one. Value, or what is good for the organism . . . is for the organism a telic end state, an intrinsic value, not always a felt preference. (257)

Of course such value is interpreted, and again teleology offers the rationale for our assuming that judgment.

Others have stretched value even further. For instance, Hans Jonas assumes that humans are the most elevated product of nature, and thus only humans are capable of deciphering and taking responsibility for the world they cohabitate (Jonas 1984). Truly, a divinely inspired Adam, for on this view, we read the law of nature and adjudicate according to our collective wisdom. Michael Serres opines similarly when he asserts:

> The life of the entire species is in our hands; it is a basis as true and faithful to things as that of the sciences themselves. We are entering a period in which morality is becoming objective. (*Le Monde*, January 21, 1992; quoted by Ferry 1995, 87)

Voila! Facts and values have been superimposed, one upon the other. Ethical claims have become objective, or perhaps more correctly, radical environmentalists have stretched the science to justify their moral advocacy. In the political domain, such moves might well be interpreted as disingenuous by those opposing the position advocated. Not taking sides, let us simply observe that (1) the invoked facts and values are entangled; and (2) the issue is not values per se, but *which* values are used and for what larger argument.

In this example, the so-called boundary question has moved from a sociological description back to a philosophical one; the same general demarcation problem identified by the logical positivists characterizes much of the debate about environmentalism. As the above comments show, ecological ethics has been supported by biological teleology. That connection is made effortlessly because of the hidden value structure of the biological description: teleology is itself an interpretation (see below). We have already discussed the failure of rigidly separating facts from values and the resulting collapse of the demarcation schema, but, at the same time, we are faced with an incorrigible mixing that leads to a weak moral argument. Indeed, must we frame environmental ethics on biology, and if not, how well can a border be placed around biology to protect its own neutrality? The following description of teleology illustrates the futility of firm demarcations, and beyond that philosophical distinction, we witness how the collapse of the fact/value dichotomy illustrates the "porosity" of the walls around objective science. Once described, we will turn to a strategy for a more honest brokering of science for moral discourse.

Teleology

Teleology gives an account of something by reference to an end or goal. Teleological explanations span the entire range of biology, from the apparent goal-directedness of embryological development, to the adaptive character of traits and organic systems, to the seeming "purpose" of behavior. The teleology hovering over both molecular descriptions, such as enzymatic cascades and complex social behaviors, is to some a "specter" of a tainted mode of thinking. Why? In answering that question, both in its historical and analytical context, we might discern more clearly the source of confusing teleological descriptions with ones that give rise to *value*.

Teleology is integral to an older natural philosophy that sought a philosophy of nature in human terms. Current biology is a science devoted to providing mechanistic explanations, and to the extent that a telos orients its theory, we witness vestiges of the older metaphysics. To put the matter simply, although we detect teleological descriptions in modern biology, they represent the inadequacies of scientific aspirations to offer *mechanical* explanations for organic processes, whether physiological, developmental, or evolutionary. Although I believe that such teleological descriptions cannot be entirely purged from biology and must continue to serve a complementary role to mechanical explanations, we must acknowledge the precise scientific status of *telos*.

Many see biology caught in a fundamental paradox. Specifically, biology is characterized by two ways of thinking: an objective program that seeks a description of biological processes, and a teleological mode of addressing those phenomena in terms of function.[6] Teleology employs descriptive criteria for elucidating seemingly intentional life processes, seemingly far removed into the future and thereby determined by causalities quite distinct from those normally encountered in characterizing biochemical and biophysical processes. "Purpose" smacks of a subjective projection or interpretation, yet notwithstanding the scruples of a science determined to find mechanical explanation (causes), biology must nevertheless rely on *telos* to order its theory and methodologies. Teleology functions as the projection of a particular rational understanding, and although potentially distorting, such descriptions orient inquiry. Let me emphasize, teleological *descriptions* do not substitute for mechanistic *explanations*, but because of the descriptive character of the life sciences, such schema are complementary to physical and chemical explanations of biological functions. This duality remains a legitimate strategy. There is no contradiction; simply, two complementary ways coexist to describe organic phenomena. The difficulty, due to the incompleteness of contemporary biology, is that we cannot truly separate the two, and that our understanding—whatever scientific erudition our biology offers us—intrinsically intertwines our biophysical and biochemical descriptions with our teleological orientation (Rosenberg 1985, 255).[7]

I submit that teleology rests on the very fault line between *fact* and its *interpretation*. Teleological descriptions are still necessary, serving as proto-theories, and thus requisite to placing the "objective" data in order. To expel teleological descriptions, and here I am referring only to descriptions of end-seeking function, would strip biology of its *logos*. The science must explain function mechanistically, that is, with factual descriptions, but those facts are placed in some functional (interpretative) edifice. There is no contradiction but a tensioned complementary between problems stated in functional terms and explanations given in mechanistic ones. Nevertheless, teleology creates an "aura" of interpretation that surrounds facts. The motivation for a purely objective account resides in the foundations of a nineteenth-century science that, in its starkest statement, attempts to give factual accounts devoid of interpretation. The attempt to purge teleology from biology in the nineteenth century was motivated by this concern, for biologists were aware of the ominous distortion that projection of an interpretation might have on the creation of facts, thereby corrupting theory construction built upon positivist ideals. On this view, teleological explanations might serve as an undeclared theory posing as fact, and in such constructions, facts situated by theory are then used to erect the supposed objective edifice to support that theory in a circular chain of reasoning. But if we allow an objective science to be judged essentially on the success of its ability to predict phenomena and cohere to explanatory principles deemed congruent with its own theoretical construct, then we may legitimately regard teleological descriptions as part of the very fabric of biological science. It serves to order observations and orient theory, which then is tested anew. And more to the point, biological phenomena simply cannot be described in isolation from end-seeking, purposeful function. The value-free ideal is shaken, for at any level, such intentionality is interpretive, even projective of human bias, and thus biology remains shackled to an epistemology that fails the most stringent positivist requirements.

By accepting objectivity and its supporting telos for the service each performs, we acknowledge their inherent limits and at the same time recognize their particular contribution to the scientific enterprise. The mistake is to extend *telos* to support some broader meaning, for that extrapolation, teleological thinking has justly been attacked. Given the power of this formulation for the *science* of life and its philosophical implications, environmental ethicists might well ponder the wisdom of grounding their own arguments on the foundation offered by *telos* and its various expressions.

From Whitehead (1925) to Daniel Dennett (1981, e.g., 28), numerous commentators have noted that the intentions we ascribe to natural systems are abstractions of a sort, in which a rationality resembling our own is ascribed to natural systems or behavior. It is this rationality that enables us to understand and to explain them using our own intelligent faculties. It reflects a deep-seated confidence in the rational order of nature, and we thus profoundly both endorse and recapitulate Kant's project. As Herbert Simon (1969) has noted, this amounts to analyzing natural systems as if they were artifacts, attributing to them the same type of rational adaptation of means to ends that we employ in the design of our machines. Beyond what Gregory Bateson (1980, e.g., 299) and others have underscored as the naïveté of believing that nature conforms to our particular scientific rationality (Atlan 1993, n. 83, p. 91), another startling philosophical revelation presents itself: this projection of our rationality upon biology "is an even stronger postulate than that of a rational intelligibility of nature, namely, that of an *intentional* rationality in nature" (75; emphasis in original).

Imposing goals on nature results from anthropomorphological reasoning. Intentionality, used to characterize psychological or social action of conscious beings, is applied to apparently intelligent behaviors adapted to the achievement of some goals observed *in* individuals, whether human or animal (Atlan 1994, 74). As illustrated, there are those who use the so-called "teleological centers of activity" (to quote Taylor [1986]) as a node of value and from there take a short step to assigning a true moral standing of life. Of course, one might easily enlist in such an ethical venture, but let us do so with a more self-aware and honest rationale. Taylor and fellow travelers make their move from a construal of biology in which their interpretation supports a metaphysical picture that has a strong grip upon our culture, one upon which the epistemological basis of teleology resides. The problem lies in large measure with the "semantic instability" of the term teleology.

In summation, teleological descriptions are, by their very nature, interpretative. In a strict biological context, they are used to define end-seeking function. The category mistake occurs when one proposes that purpose has moral standing. Certainly human intentionality is structured by a value system: we make choices embedded in an ethical construct. But do purposeful animals have moral standing? The ecological ethicist must believe that beyond sharing purpose, animals share with humans the same biological origins of morality itself, the very matrix by which purpose, whether human or animal, attains its moral meaning.[8] Thus telos effortlessly jumps from human to animal morality, from which an ethical framework, purportedly built from science, is erected. Of course, as seen in this case example, the lines of division between the categories we call science and subjective values remain blurred. Indeed, in this instance, we have a particularly rich interplay that we can discern in the historical development of environmentalism as a distinct response to the challenge of industrialization and the economic effects of mass population growth in the nineteenth century.

Environmentalism's Religious Origins

The division between scientific description and moral argument obviously has been blurred in creating the conceptual framework for environmental ethics. This intellectual structure arose from a complex historical development and diverse cultural forces. Of these, perhaps the most important are what must be assigned as religious motivations. Indeed, twentieth-century environmentalism is rooted in the romantic period, when new spiritual sentiments celebrated the sublime natural as a focus of religious awe and wonder. The moral structure of the contemporary Green movement expresses this deeper spirituality.

The romantics' signal achievement was to move beyond the strictures of reason devoid of subjective ways of knowing. They dispensed with the effort to unify reason or to bridge a divide between humans and nature, but instead steadfastly held to a new principle: rather than residing above nature akin to angels with moral responsibility and self-consciousness, humans live in nature as essentially wild and natural. Correspondence between natural law and God's law allowed the study of nature to provide key moral and spiritual insights, and further, the source of human virtue would be discovered, not by revelation, but through communion with nature (Emerson 1971). Accordingly, any of the social constraints on human appreciation of the natural distracts, and potentially corrupts, one's exploration of his or her essential character and goodness. In short, from Rousseau to Thoreau, the *wild* endows the vital center of human creativity and human virtue with divine authority and inspiration (Tauber 2001, 177–87).

In asserting that the natural is the human's core being, the romantic then makes his mission the discovery (and resultant intimacy) with a primitive essence, which connects humans with the cosmos. The wild, because of its very character, cannot be "known," that is, tamed or rationalized, and made a species of consciousness. All those modes of knowing that we must pursue are sorry residues of a primary knowing. In the wild, reason does not rule; it can, at best, only mediate. Nature is processed by the mind (and thereby constructed [Evernden 1992]). Hence, nature is experienced in myriad variants as a *derivative* product of "the wild." This residual experience must be overcome as both an epistemological endeavor and a moral one. In this vein, the arch American environmentalist, Thoreau, wrote:

> I wish to speak a word for Nature, for absolute freedom and wildness, as contrasted with a freedom and culture merely civil,—to regard man as an inhabitant, or a part and parcel of Nature, rather than a member of society, I wish to make an extreme statement. (1980, 93)

By tapping into his own "wildness," Thoreau believed that nature was not redeemed so much as transformed in an ongoing creation of his imagination. So now we encounter Thoreau for the second time in this essay, moving him from his general effort to integrate his science with personal accounting of the natural world to the specific modality of environmentalism. In that move, he offered a new sensibility. His vision challenged each of us to create a unique and profound spiritual relationship with nature.

In championing the individualism for which he is most famous, Thoreau promoted an intimacy with nature, which in many respects complemented a cultural ethos. After all, he lived in an agrarian society, and Concord's farmers can hardly be regarded as divorced from their natural environment, which held virtually all fortunes. He was sensitive to the Native American heritage, and like other American romantics, regarded those original inhabitants with a certain degree of awe. And when placed within the tradition of American exploration, Thoreau shared a reverence for natural beauty already held by those searching for new trade routes (as in the case of Lewis and Clark) or studying the fauna of a new land (for example William Bartram or John James Audubon). Thoreau's place as an architect of our own environmental movement lies elsewhere.

Thoreau's environmentalism draws from a religious sensibility (Hodder 2001), which combines two components: first, the notion of the sublime coupled to a crude pantheism, and second, an acute self-consciousness, which focused his own nature study and worship as an exercise in identifying himself in relation to nature. Together, a metaphysics of nature and a metaphysics of selfhood combined to offer a particular vision of humans *in* nature. That project helped form our own environmentalism. Many have been inspired by Thoreau's example, and twentieth-century environmentalists have hailed him as their prophet. Proclaiming an anthem to the simple life, the common laborer, the goodness of man in nature, Thoreau, the American Rousseau, took the moral high ground in a society adjusting to the challenges of switching from an agrarian culture to an industrial one. He voiced what became a modern chorus of criticisms of mass society's commercialism and depersonalization. In celebrating the wild as a value *sui generis*, he made a moral claim against his peers, whom he regarded as corrupted by a social mercantilism and a dehumanizing positivism.

Thoreau's importance as social reformer and an early spokesman of environmentalism resides more deeply on his metaphysics of the self. His was a particular response to a common romantic query, namely the problem of self-consciousness that will not abide any respite from its own relentless self-scrutiny (Tauber 2001). Thoreau offered the antidote of self-knowledge and self-reliance proclaimed at the end of *Walden*. In his admonition, he advised an acknowledgment—"We know not where we are" (1971, 332)—and then demanded establishing the wherewithal to chart one's course to "find ourselves and realize where we are and the infinite extent of our relations" (171). In the nineteenth century, this optimistic message that Americans were masters of their own destiny and thus capable of conquering both a continent and themselves signified a trust in human resourcefulness—spiritual, existential, and psychological. The triumph of a pioneer civilization, one which celebrated God's bounty and goodness, faced a crisis with the closing of the frontier. In a surge of national self-consciousness, the environmental movement was born to celebrate God's splendor. Thoreau presciently marshaled this spiritual buoyancy and made it the foundation of a new nature religion. On this view, he was a psalmist of American success and optimism, and for millions of Americans, he is a prophet (Tauber 2001, 219–21; 2003). When he wrote in *Walden*, "Everything is startlingly moral," Thoreau understood that nature has meaning and significance *for* humans and how such value projected upon nature served spiritual needs. After all, for Thoreau, Nature was a source of vitality; an origin of spiritual recognition and union with the Unknown; the source of all beauty; and finally, the salvation of corrupt materialists.

Thoreau's relation with nature arose from an existential sense of an isolated ego, where nature resided "out there," ultimately magnificent beyond human imagination, but alien, strange, and potentially dangerous. To be sure, his self-consciousness, the awareness of his separation from nature, drove much of Thoreau's efforts to join nature more intimately. That irredeemable separation between himself and the world in which he lived remained an abiding problem, and as we peer at Thoreau through these spectacles, we see that environmentalism presents a response to a deep metaphysical disjointedness. In our own era, when the environment teeters in precarious balance with economically driven assaults, the environmentalist credo has taken on a new urgency and compelling rationale. But beyond that immediate crisis, looking at Thoreau as a romantic, one alienated and disjointed from the world, we see how an attempt to better integrate himself with nature offered respite from an existential anxiety. So at the heart of Thoreau's environmentalism, and our own, we find a primitive motive for what is fundamentally a religious response to alienation and danger.[9] The intuition of nature's sanctity then addresses a primordial human experience: honor the gods and revere them or be smitten by their wrath.

The irony of employing a *scientific* rationale for this essentially *religious or metaphysical* problem only highlights how science has become so pervasive to Western thought that its offering must somehow be incorporated into a venture residing well beyond the claims of objective knowledge. My assessment points in several directions, but the most important concerns science's place in constructing our *personal* world. This humanist project draws from many sources, and in the conclusion of this book, we will explore the challenge of integrating the objective with the subjective to bridge domains that seem separate but flow freely one into the other. We have emphasized the tribulations of an insular objectivity operating with a reason onto itself. Now we ponder how the product of objective science might become personally meaningful.

Conclusion

The Challenge of Coherence

Meaning is wider in scope as well as more precious in value than is truth.

John Dewey, "Philosophy and Civilization"

We began this essay characterizing science as an epistemology and outlining its place in contemporary American culture. In the clash of metaphysical views accentuated by the Dover case about intelligent design, a courtroom battle waged over an argument concerning reason, which was settled by law, not compromise. Indeed, metaphysical claims are not adjudicated: differing visions simply slide past each other with little traction and correspondingly little influence exerted on other contending positions. Preferring to eschew such debate, science rests on its practical achievements and thereby makes only modest metaphysical claims. Yet it is not so easily excused from such discussions, for its original, ancient agenda continues to beckon—the metaphysical wonder that drives scientific inquiry and the findings that must be translated into human meaning. (Here I am referring to meaning in the humanistic context, not "meaning" in the sense explored by twentieth-century analytical philosophy [Soames 2003].)

In the modern period, the Cartesian construction of the mind/nature divide established the terms of philosophical discourse. Early modern epistemology sought to discern the nature of human perception and the ability to derive mental "pictures" of the world. Science, with its logic and universal methods, did offer a powerful model for understanding how those sensory findings are extended into scientific facts and laws, a project Descartes thought would result in the axiomatization of nature. That dream was abruptly interrupted by Hume, who argued that causality was merely a psychological habit (or conceit), as opposed to a logical or inherent property of nature. This apparent artifice of mind awoke Kant from his "dogmatic slumber"; his entire transcendental project was designed to provide the conditions by which mind addressed the quandary Hume presented.

Kant posited that *because* of reason's autonomy, the mind became the "lawgiver" to nature (i.e., it provided law as a product of human cognition and imagination), and, at the same time, mind patrolled and created its own human social and spiritual universe with a reason designated for that purpose (Neiman 1994). Kant thus directly confronted the human/nature divide with reason's own "division," and his formulation then demanded some understanding to reunify that which had been split. He failed, and much of German Idealism in combating the subjectivism he bequeathed attempted to resolve the imbroglio (Beiser 2002). But that project also floundered, and with the ascendancy of positivism in the mid-nineteenth century, as already discussed in several contexts, the metaphysical challenge became subordinate to other concerns.

Science, left to its own narrow pursuits, ascended to great heights of technical mastery of nature, which largely displaced the Cartesian-Kantian question from serious philosophical discussion as positivist philosophy assumed its own empirical conditions of knowing. To the extent that the issue was entertained, it became, in the romantic formulation, how to place the self in relation to the nature it experienced. While the objective component of that problem was soon adopted by a new discipline, psychology, the existential expression of the deeper metaphysical challenge soon dominated discussion of the signification of science's findings. In other words, with objectivity given, the unresolved problem became *subjectivity*. And so, any attempt to employ science as an instrument for *unifying* reason would have moved against a strong positivist tide that was firmly committed to its other concerns. However, the unification problem reappeared in the early twentieth century, perhaps in different guise, yet nevertheless demanding attention. Whereas the nineteenth century posed the subject/object divide in opposition, some twentieth-century philosophers still sought a synthesis. And so we come full circle back to the challenge of finding coherence between the world science describes and the personal universe in which we each live.

Of course, one might argue that "coherence" is neither necessary nor possible in a postmodern world. Much philosophical and cultural criticism commends that view, but I reject it. I assume a dissenting position on the basis that, although incoherence may characterize our lives, such disjointedness is a symptom of a metaphysical crisis, not a *necessary* result of our post-industrial existential condition. Accepting the description of our predicament does not result in some required acquiescence that we must abandon efforts to redefine our position. Incoherence is not the "steady state" of postmodern ennui; *overcoming* our predicament is. Indeed, if Nietzsche diagnosed nihilism, he also prescribed its antidote. I follow his lead here. I will not argue my position through an exercise of logic (although I could, considering the importance of coherence as an epistemological criterion of truth); instead, my orientation is based on a *moral* appraisal of who we are and what we might become. Thus I am making a choice: I *select* a way of "being-in-the-world." On the integrative view, science shifts from its role as a wedge between humans and nature to an instrument that helps to bring us *into* the world. This issue has a long and complex intellectual history, which we will now briefly review.

Science as a Worldview

In our own postpositivist age, the romantic complaint has still not been answered. The "spectator gap" remains as efforts to humanize science by bringing the objective world into closer proximity to the knowing subject have yet to yield a full harvest. Indeed, the very division of self and other makes unity *the* problem, given the power of those instruments that have so successfully conspired to separate and divide. Nineteenth-century romantics voiced their own response, while the sentiment, if not the same arguments, have resurfaced at various points in the twentieth century in the philosophies of Heidegger and Husserl, the environmentalism emerging from American Transcendentalism (Tauber 2003), and the various antiscience critics of the 1960s (e.g., Mishan 1967; Roszak 1972). Each flailed against an invidious scientism, and while holding to their respective secular course, each attempted to find a passage to escape the compromise of reason's full expression. In the recent Science Wars, this resistance took the form of rejecting the claims for science's unique epistemological standing.

Pulling science back from its steadfast aloof observation (and mastery) *of* nature to include its complementary original pursuit of offering a platform on which we might seek meanings and significations to humans *in* nature highlights scientific knowledge engaged in a larger humane project. At the very least, some still seek a science that moves beyond mastery of nature to participate in a more encompassing *Weltanschauung*. That requires reconfiguring the knowing agent within a revised philosophical complementarity—outside observer/integrated participant—and discerning the values that would allow such a coupling. I believe that those synthetic values lie dormant in the scientific enterprise, and we may now consider them in new light.

Others have also called for a "return to reason" (Toulmin 2001)—a broad reason that allows for different kinds of discourse with different standards of knowledge to capture a spectrum of experience directed at different ends. This pluralistic, pragmatic reason is designed not only to identify the dilemma of placing science in a humane perspective, but also to ask the key question, How might reason serve various kinds of creativity to make the world more meaningful? Humanism frames that question and thereby ultimately brings science back into its fold simply by posing that challenge. This humanism is not necessarily the humanism of liberal discourse, secularism in its various guises, nor even science as a branch of philosophy—each of which commands interest in this discussion—but rather is a particular way of regarding science as a *human-centered* activity.

Such a project has two major components. The first is to weave a web of beliefs that places scientific knowledge conjoined with other understandings. Thus, beyond the material fruits of scientific labor, the most profound effect that demands attention concerns science's worldview, or as Heidegger noted (1954a/1977), acknowledging that there *is* a worldview at all! The theories and methods that have demonstrated the worlds of molecular biology, tectonic plates, quantum mechanics, and so on have markedly altered how we conceive the world in which we live and our relation to it. Further, the human sciences, for better and for worse, have bestowed their own theories on human character and conduct. This set of issues falls under the rubric of "science and ethics" or "science and policy," matters we have already considered.

A second dimension of integration comprises the family of problems that may be called the *romantic awakening*. These collectively arise from science's instantiation of the Cartesian subject/object divide, which places the observer outside the world to peer at it and to peer at herself. Always aware of separation, and appropriately so, since science would purge itself of subjective contamination, this "subjectless subject" (Fox Keller 1994) faces the metaphysical challenge of finding herself in the world described without her. This refers to shifting the human "stare" *at* the world to human placement *within* it. Simply, our metaphysics poses the challenge of how to mend the world, to make the world—humans and nature—whole again. As discussed in the previous chapter, one expression of this sentiment may be found in environmentalism, which draws from both the earth and biological sciences, as well as from a religious sentiment (Albanese 1991; Dunlap 2005).

Twentieth-century continental philosophers, most notably Heidegger (1954a/1977; 1954b/1977), Weber (1919/1946), Husserl (1935/1970), Sartre (1943/1956), and Gadamer (1981) taking their lead from Goethe (1794/1988) and Schiller (1801/1993; Beiser 2005), repeatedly addressed this metaphysical quandary and provided commentaries about a reality depicted objectively, that is, a world in which humans self-consciously reside separated from that world. From their descriptions, the challenges of defining meaning and significance of human existence took diverse courses, but linking them is the overriding concern of finding a philosophical means by which to stitch together the subject and its object. A scientific worldview must be translated into terms of human meaning, and while meaning was traditionally offered in religious terms, other options exist. To arrive at such opportunities, science must be placed within its widest philosophical calling, and this requires understanding science as part of the humanistic tradition.

For better or for worse, Heidegger set the argument in the starkest terms by attacking the very basis of objectification and the pursuit of what he referred to as "object-ness." Beyond the common concern that an imperialistic science may unilaterally establish *the* picture of the world, he perceived that "world picture . . . does not mean a picture of the world but the world conceived and grasped as a picture" (1954a/1977, 129). (Again, the isolated observing Cartesian subject is the pivotal philosophical turning point upon which the entire enterprise rests.) However, in Heidegger's view, irrespective of the potential freedom the inquiring subject might now possess, science's domination was ultimately limited because the method of science, its very essence, excludes that which cannot be incorporated into its world picture, namely, that which cannot be represented and directly related to humans in the anthropomorphic cosmic picture. So, even beyond the overwhelming presence of science in industry, education, politics, and warfare—indeed its power to influence or shape every dominion of our lives—science has become "*the theory of the real*" (1954b/1977, 157; emphasis in original), and as such, according to Heidegger, the scientific picture restricts human understanding within narrow boundaries. A scientific worldview cannot make a totalizing claim to knowledge, and from his perspective, we have been left unable to probe the shadows we recognize beyond the borders of such inquiry. That access is limited to only one way in which nature exhibits itself, or in other words, nature is not fully captured by objectifying the world in measurement, the Galilean criterion of validity.

Heidegger's views converged with others who generally agreed on the limitations of the then current expectations of science to provide a comprehensive worldview and a basis by which knowledge might be unified under its auspices, for example, Whitehead (1925), as well as Husserl, who dramatically posed the challenge in *The Crisis of the European Sciences*:

> Merely fact-minded sciences make merely fact-minded people. . . . Scientific, objective truth is exclusively a matter of establishing what the world, the physical as well as the spiritual world, is in *fact*. But can the world, and human existence in it, truthfully have a meaning if *the* sciences recognize as true only what is objectively established in this fashion. (1935/1970, 6–7)

What began as Descartes' Dream, became Husserl's nightmare; a philosophy that sought to describe nature in formal terms (i.e., geometrically or mathematically) has left science as "a residual concept" (9). By this, Husserl noted that "metaphysical" or "philosophical" problems that should still be broadly linked to science under the rubric of rational inquiry were now separated over the criterion of fact. In a powerful sense, "positivism . . . decapitates philosophy" (9) by legitimizing one form of knowledge at the expense of another. So despite Kant's own depiction of theoretical and practical reason united under a common regulative banner, that division was never truly mended. The "crisis" bestowed by the sciences resides in their separation from their original philosophical foundations (Hopp 2008). This resulted in divided rationalities and a corresponding inability to address human interests as defined in a humanistic framework, or as Husserl put it, "in this vital state of need . . . this science has nothing to say to us" (1935/1970, 6).

Husserl's criticism confronts positivism in terms quite divorced from any technological influence exerted on the wider social domain. He drew the distinction between science, a product of reason, and reason in its existential and metaphysical roles (the vital source of scientific inquiry) to reiterate the original romantic complaint.[1] Husserl called for a coherent reason, a means of unifying experience in its various expressions, so the "crisis" was presented as a reflection of the deeply divisive nature of two kinds of knowledge. In seeking to establish a common philosophical grounding for each sphere of experience, Husserl revisited the "unity of reason" problem bequeathed by Kant (Harvey 1989; Neiman 1994). Furthermore, for Husserl, the crisis was not limited to science or philosophy but reflected a fundamental challenge to European cultural life, its total *Existenz*.

Many tributaries have fed into the abandonment of a unifying universal philosophy. To the Anglo American ear, such speculation seems not only foreign, but strangely whimsical and even effete. For this dubious group, scientific reason is assigned to govern one domain of knowledge, and other kinds of reason are left to matters of value and ethics. Indeed, lines have been drawn precisely on this basis, and those who discard the very possibility of some enveloping philosophy basically ignore Husserl's project or dismiss it as misconceived. For those in that rejecting camp, "multifocal" reason characterizes human life, and to pursue Husserl's project smacks of eclipsed metaphysics. Indeed, the twentieth-century philosophies attempting the Husserlian enterprise—existentialism, Marxism, structuralism, Heideggarian phenomenology—have each proven incapable of the task assigned themselves. Instead, following Wittgenstein, analytic philosophers have sought to show that the very conception of such a venture is misconstrued. Ironically, this general posture may well be the most enduring of the contributions made by the logical positivists, for while they failed to formalize science, they succeeded in discrediting projects such as Husserl's. I take this division as given: an irreconcilable difference in the Wittgensteinian and Heideggarian competing conceptions of contemporary philosophy translates into divergent intellectual aspirations and different philosophical expectations. However, if we assume a pragmatist orientation, one directed at the human use of science, perhaps a more satisfying unification might develop.[2]

Bridging the Divide

The need for finding meaning, this relic of an ancient metaphysics, resides deeply within human psychology, and science is hardly immune from being co-opted for this larger purpose. Indeed, if scientific knowledge has become a paragon of truth and a source of wonder at nature's order and function, how could those findings remain immune from being placed within a larger context? Henri Atlan suggests that this metaphysical posture results from a profound desire for science to provide a comprehensive explanation of nature. He refers to this as a mystical aspiration:

> The need for an explanation of reality is, fundamentally, antiscientific. The satisfactory explanation is a bonus, the esthetic pinnacle that accompanies and sometimes completes . . . the result truly sought; technical performance . . . For the practitioners of contemporary science . . . the need for explanation is merely a relic of metaphysical, indeed religious, wonder. (Atlan 1993, 193)[3]

Following this theme, Gaston Bachelard (1934/1984), rather than lamenting the contamination of such a metaphysical remnant, celebrates its role. He saw in the pursuit of meaning the motive force of research, one that would animate scientific query in a twofold fashion: nature not only has a rationality which invites discovery (and thus enables humans to place themselves within nature from which objectivity separates them), but more intimately, that knowledge, translated into wonder, provides the emotional recognition to marvel, and thus regain, a lost enchantment. To find personal meaning represents the process by which objectivity and subjectivity (both acknowledged and justified) are brought into proximity, to overlap, and even to integrate. To speak of nature, we draw from both the objective scientific accounts as well as the relational aspects derived from the pervasive metaphysical picture science presents. The pursuit of the real, in the end, is a quest for *meaning*. In this latter task, we ultimately define and place ourselves within the cosmos.

To shun metaphysics does not mean we escape its call.[4] The metaphysics of theology may not beckon, but metaphysical wonder remains, and even more deeply, the task of understanding the existential placement of humans in the world cannot be ignored. In the hegemonic triumph of a reality composed by science, we may have exchanged one set of beliefs with another, but that does not signify the absence of a metaphysics that helps define our understanding of the real. Rather than deny the metaphysics of this scientific age, perhaps we should delve more deeply to understand them. I believe an affirmative answer must prevail because, while we might resist alluding to metaphysics in this "post-metaphysical" era, we cannot escape the question of *reality* and our place in it. And defining that reality extends far beyond the purview of science and its various conjugates. If philosophy cannot address the challenge, other venues will continue to offer their means of expression—art, religion, literature, music, political discourse, and so forth. While each can proceed on its own, historically, philosophy has been instrumental in defining the central metaphysical questions. I cannot fathom philosophers abdicating that role. If my essay has a fundamental theme, it is to recall that ancient mission.

In the romantics' resistance of the mighty positivist tide, they enjoined a metaphysics that rejected the narrow picture of reality science offered.[5] The romantic disenchantment resides at several levels but may be traced to a failure: Descartes' dream of formalizing nature to mathematics and thereby discerning the divine cosmic machine failed. The promise of defining nature elevates science as the bearer of all *seemingly* esoteric knowledge, but because the process of intellectualization denies metaphysical or religious enchantment, the problem of the Cartesian *cogito*, separated irreparably from the world it surveys, leaves the subject alienated. On this view, science cannot address the problem of meaning. Each individual is left to seek value on his or her own, and as a consequence, nihilism raises its hoary head. More saliently, from the romantic perspective, science is reciprocally rejected by a temperament that finds the radically objective worldview bereft of personal and, more importantly, emotional meaning:

> When I heard the learn'd astronomer,
> When the proofs, the Figures, were ranged in
> Columns before me,
> When I was shown the charts and diagrams, to add,
> Divide, and measure them.
> When I sitting heard the astronomer where he
> Lectured with much applause in the lecture room.
> How soon unaccountable I became tired and sick.

> Till rising and gliding out I wander'd off by myself,
> In the mystical moist night-air, and from time to time,
> Look'd up in perfect silence at the stars. (Whitman 1865/1902)

Walt Whitman's pronouncement was, and continues to be, a widely shared sentiment. It is commonplace to note that, since the romantic period, the analytic, mechanical, abstract qualities of science have displaced the primary encounter with nature, leaving the personal, emotional, and aesthetic dimensions of experience to find their own course.

According to these romantics, the picture of reality offered by science is hardly "real" in human terms. Any picture presented must be translated into significance. Thus, beyond the knowledge derived from scientific inquiry, aesthetic and spiritual insight about the splendor, wonder, and miracle of existence must also be acknowledged. Whereas the scientific facts themselves cannot carry such understandings, meaning is projected by the humane appreciation of the facts provided. In this sense, the reality science provides is ironic, inasmuch as the objective picture is hardly the reality we know in any intimate or personal terms. From the quantum universe to astrophysics, from the human genome to tectonic plates, scientific explanation is divorced from everyday experience. Here, at the interface of the objective/subjective divide, we again witness the radical distinction of fact and value collapse.

Some kinds of knowledge are produced at the far end of the objective pole, and such knowledge demands (appropriately) minimal "contamination" with the subjective. But this is not the issue, for that battle has been long fought and decided. The point pursued here is something else entirely: personal knowledge is not necessarily contingent or arbitrary. Rather, to translate an objective picture of the world into terms that have human meaning and significance requires other kinds of values and assessments, and these too have their legitimacy and just applications. The issue is not to entirely purge the subjective, but to recognize its rightful place in the tribunal of judgment, where knowledge is ultimately valued and deployed for human use and understanding; for it is both useless and irrelevant divorced from the reality of the personal domain. In short, Knowledge is inexorably valued along the *entire* objective/subjective continuum.

Those who would discard romanticism's yearnings maintained that "meaning" was never listed on science's menu, at least not as a main course. Whatever meaning is derived from scientific findings must be taken à la carte, probably only as dessert. Accordingly, meaning comes from outside of science, and such interpretation arises as a matter of choice, a question of belief and personal need. Given that both parties agree that meaning resides beyond science, it seems that the romantic quest remains for those so inclined. Simply, let those who seek a better synthesis carry on as best they can. Before accepting this de facto conclusion, let me restate the "unifiers" case.

A Moral-epistemology

Let us return to where the path of science veered off from its parent philosophy. As we have discussed, beyond its unique object of study and a particular empiricist methodology to fulfill its goals, science placed particular values on certain kinds of knowledge and dismissed others, and rightly so in light of its own pursuits. However, in following its own way, science left no means to integrate its own agenda with the humanistic body it deserted. Accordingly, as the romantics so well understood, *meaning* must be sought in some fusion of the personal domain with objective knowledge of nature. In seeking a synthesis wherein objectivity might be mixed with the personal and thereby achieve what Polanyi called "personal knowledge" (1962), certain cognitive strategies might be employed. We reviewed them earlier, but essentially I have presented how both science and *poesis* should be viewed as vehicles of knowing or expressing. One might assign them the status of "tools" of an interpretative faculty that confers *human significance*. Neither science nor poetry individually suffice because, in a sense, they are appendages to a deeper cognitive function, a faculty Kant called "judgment." From this perspective, meaning cannot directly arise from epistemology or any of its branches, but rather arises from a dynamic synthesis—the moral vision (or orientation) of the knower who weaves facts into their fabric of signification. If the positivist insists that science reside within narrow epistemological constraints of "knowing," then the larger picture of the world remains incomplete. Alternatively, the true scientist will not leave the world bereft of his own presence. One would not place the subjective into science and contaminate its hard-won reality; rather, we seek a way to add the personal to make science fully real . . . in human terms. In short, the humanist rejects the positivist stance, not so much because it distorts reality, but because it captures too little:

> The future will no doubt be a more natural life than this. We shall be acquainted and shall use flowers and stars, and sun and moon, and occupy this nature which now stands over and around us. We shall reach up to the stars and pluck fruit from many parts of the universe. We shall purely use the earth and not abuse it—God is in the breeze and whispering leaves and we shall then hear him. We live in the midst of all beauty and grandeur that was ever described or conceived. (Thoreau 1981, 460)

So I again turn to Thoreau to illustrate this expansive description of science.

Thoreau sought "facts," the lexicon for his descriptions; indeed, they were the métier of his life (as he exclaims shortly after moving into his Walden cabin: "I wish to meet the facts of life . . . face to face" [1984, 156]), and he would engage them "directly." Concomitantly, he also recognized how facts are contextualized by their standing within a certain milieu of conceptual support and theoretical understanding. Thus he appreciated how facts carry the very complexity they are used to simplify. Note how Thoreau portrayed the knower as processing facts to attain their full significance and meaning:

> See not with the eye of science—which is barren—nor of youthful poetry which is impotent. But taste the world. & digest it. It would seem as if things got said but rarely & by chance—As you *see* so at length will you *say*. When facts are seen superficially they are seen as they lie in relation to certain institution's perchance. But I would have them expressed as more deeply seen with deeper references.—so that the hearer or reader cannot recognize them or apprehend their significance from the platform of common life—but it will be necessary that he be in a sense translated in order to understand them. (1992, 158; emphasis in original)

The intimation underlying this passage from Thoreau's *Journal* describes the role of the observer in signifying the world.

From Thoreau's perspective, the scientific view of the detached observer should extend objective findings gleaned from nature to other domains of individual experience—the emotional, the subjective. The problem for him was not how the scientist attempts to be objective, but rather how knowledge of the world becomes *signified*. The import of that synthesis concerns the very nature of objectified knowledge itself, which is thereby transformed from its public domain back to the individual. In dismissing a positivist notion of objectivity, Thoreau claimed that we must make choices and thereby assign particular importance to one kind of observation over another. We weigh information; certain details become important within the context in which they are seen; the observer creates that context. So, on the one hand, Thoreau sought to "capture" nature through meticulous observation; on the other hand, he "personalized" reality so distilled by situating himself in the order and beauty of the natural world (Peck 1990; Tauber 2001). The aesthetic and spiritual were not *attached* to objective seeing, but were rather *constitutive* to it. Simply, for him, poetry and art were cognitive faculties to meaningful order experience (Croce 1902/1972).[6]

For Thoreau, this processing of facts yields a global picture that, because of its source deep within the knowing self, saturates his being: "the truth respecting *his* things shall naturally exhale from a man like the odor of the muskrat from the coat of a trapper" (1992, 158; emphasis in original). Note the emphasized possessive pronoun, *his*. Thoreau has personalized knowledge to strike "truth," which, rather than residing out there in the world, attains its status only through the capacity of a sensitive and ready knower. Truth for Thoreau is not judged through some positivist standard, but rather falls squarely in the personal realm as a product of synthesis *between* mind (sensory and aesthetic faculties) and the world. So he is not creating truth by an idealistic ploy. The world, following Kant, exists as real and knowable, but a dialectic plays between the subject and his object of inquiry. Ultimately, the observing eye must gather facts, but the mind orders them into a formulation. This may be done under the formal structures of either science or poetry, but these too are translations.[7] For Thoreau, nature ultimately must be experienced, directly and intimately. As he wrote in his journal,

> I think the man of science makes this mistake, and the mass of mankind along with him: that you should coolly give your chief attention to the phenomenon which excites you as something independent on you, and not as it is related to you. The important fact is its effect on me. He thinks that I have no business to see anything else but just what he defines the rainbow to be, but . . . it is the subject of the vision, the truth alone that concerns me. The philosopher for whom rainbows, etc., can be explained away never saw them. With regard to such objects, I find it is not they themselves (with which men of science deal) that concern me; the point of interest is somewhere *between* me and the objects. (1962, 10:164–65; emphasis in original)

Thoreau endeavored to capture a humanized worldview, one which employed the fruits of scientific investigation to create his *own* world. In this sense, I regard him as an exemplar of a scientist guided by a humanistic ethos (expanded in Tauber 2001).

Thoreau offered no apology that his natural histories (spliced with poetics and metaphors) might not conform to the standards of *bona fide* scientific knowledge. Some passages did; others did not. But for him, the work was synthetic, an ongoing negotiation of melding the objective world with personal experience and understanding. So while he sought objective knowledge, Thoreau engaged a larger project. The "ethics" of *understanding* drove Thoreau's quest to know. His moral venture was a mandate of seeing—one developed, exercised, and pursued as an act of defining himself and his place in the cosmos. (*Seeing*, of course, not only included poetic visions or waking dreams, but meticulous observation.) Knowledge remained in the employ of that agenda, but science was only one of several tools to achieve the greater purpose of human-centered knowledge. He dismissed a false choice—knowledge *or* romance, science *or* poetry. Both modes of experience were crucial in fulfilling his quest for reality. Unified reason bequeathed unified experience.

In sum, Thoreau was cognizant of the limits of his role as epistemologist as one who could know the world as something real beyond his own vision. And, as a critical corollary, there were various faculties of seeing, each capturing an aspect of reality that must be forged into some unity through the mind of the beholder. So seeing was ultimately dependent on the individual's *ability* to perceive and integrate, and thus the world as known emerges in creative effort. In the pursuit of nature, the project itself was informed by a self-conscious *doing* of those inquiries. In some sense, he anticipated the later pragmatists' emphasis on the centrality of engaging lived experience. In this sense, the act of creation itself was a virtue. Thoreau's science thus became a *poesis*:

> Nature has looked uncommonly bare & dry to me for a day or two. With our senses applied to the surrounding world we are reading our own physical & corresponding moral revolutions. Nature was so shallow all at once I did not know what had attracted me all my life. I was therefore encouraged when going through a field this evening, I was unexpectedly struck with the beauty of an apple tree—The perception of beauty is a moral test. (1997, 120)

This journal entry (June 21, 1852) resounds with *Walden*'s declaration that "Our whole life is startlingly moral" (1971, 218), which may be understood as our placement on a set of coordinates defined by several axes. Just as space is geometrically defined by three vectors in simple geometry, so too might we draw a "space" by "vectors" which will analogously define the coordinates of Thoreau's experience: the first, the aesthetic imagination; the second, the imperative of attention; the third, the psychology of self-awareness. Their meeting at the origin of the vectors that delineate this metaphorical space is joined by a value-laden consciousness that guides each faculty as it probes its intention. Here we find scientific thinking participating as one faculty among several to help form that composite we call "the world."

For Thoreau, that world was signified relative to himself through a "moral-epistemology," one he forthrightly declared as a young man: "How tremendously moral is our life. After all no man can be said to live much in the senses, but every moment is the product of so much character. What painters of scenery we are. We impart to the landscape the perfect colors of our mind" (1981, 466–67). Nature could only become meaningful as knowledge of it became integrated into Thoreau's own experience. And experience, at least for Thoreau, was an imaginative and intimate encounter. To repeat the epigraph of this book:

> The true man of science will have a rare Indian wisdom—and will know nature better by his finer organization. He will smell, taste, see, hear, feel, better than other men. His will be a deeper and finer experience. We do not learn by inference and deduction, and the application of mathematics to philosophy but by direct intercourse. It is with science as with ethics—we cannot know truth by method and contrivance—the Baconian is as false as any other method. The most scientific should be the healthiest man. ([*Journal*, October 11, 1840], 1981, 187)

The world was thus made whole again as the scientific pursuit took its rightful place in the quest of reality, a reality of Thoreau's own creative making. The world, just as Emerson had pronounced at the end of *Nature*, would be built according to one's own dictates, directed by an inner compass: "Every spirit builds itself a house; and beyond its house, a world; and beyond its world, a heaven. . . . Build therefore your own world. As fast as you conform your life to the pure idea in your mind, that will unfold its great proportions" (Emerson 1971, 44–45).

To make the ordinary extraordinary became the moral endeavor Thoreau set himself. In consequence, he possessed a unique view of nature, essentially creating, by means of an astute sense of detail and a poetic eye, the world in which he lived. This Thoreau accomplished as an act of will, asserting the primacy of his own knowing. *What* he saw was determined by *how* he saw and the value bestowed on the object of scrutiny. To see the world as beautiful and spiritual, Thoreau placed lenses of enhanced sensibility before his eyes, both to focus his sight and filter it. Thus the very act of observing became a moral test of his values and his ability to live by them. He essentially composed nature in a personal format, taking what he required to make a picture of the world and of himself within it. The individuality he espoused was the *sine qua non* of the entire project. In short, instead of objectivity's "view from nowhere," Thoreau proclaimed the primacy of precisely his own vision.

Thoreau's reconfigured epistemology shifts the scientific question from matters of fact to matters of interpretation. And finding significance draws upon emotional and moral faculties quite distinct from those employed for scientific investigation, at least in a first order way. Any significance humans derive from nature or apply to nature represents a set of projected human values and interpretations. And some would join me (e.g., Bachelard, Atlan) to argue that this metaphysical assignment, not mastery of nature, defines science's deeper "end game." So while objective knowledge is obviously useful as an instrument towards generating industry (broadly understood), the romantics repeatedly demanded that scientific insight must serve to *make the natural world significant in human terms*. Accordingly, the web of beliefs in which objective knowledge, knowledge independent of individual perspective, must be integrated is a complex network of psychological and cultural elements. In the simplest of terms, science offers a powerful means of achieving control of nature, but those successes must be coupled to furthering human goals and enlarging humane understanding. If that synthesis fails or is simply abandoned, then science's full potential cannot be realized.

The Search for Meaning

If we remain committed to understanding how scientific knowledge becomes meaningful, not in its technical applications, or even as statement of fact, but rather how the picture science offers of reality becomes incorporated into belief systems and insight, then the world science reveals may be appreciated aesthetically, and spiritually. On this view, knowledge is valuable not only for its application, but also for providing orientation into a complex, often hostile, sometimes mysterious world. This latter pursuit points to a deeper stratum of experience, the one that initiated scientific inquiry in the first place: metaphysical wonder. One might describe the glorious creation of the divine, or more neutrally, translate the emotional appreciation of nature's wonder in aesthetic terms, or more simply, the "aha" of insight. Here the scientific discovery or theory is framed by humane intuitions. This attitude points back to the romantics, whose own agenda remains unfulfilled but whose challenge, I maintain, cannot be denied. How we fulfill their insight that a fractured world must be mended remains our own work.

To portray a "disenchanted" universe as the inevitable product of scientific inquiry seems to me misguided. Such an indictment cannot be dropped at the doorstep of science. Rather it rests more appropriately in another, broader context, namely finding human meaning in a secularized world, whose values are no longer divinely revealed but instead arise within a socio-political culture in answer to the needs of the historical moment. The program that might integrate the scientific worldview with human experience and interpretation involves many players, of which science is but one, albeit central. The orientation developed here points to a philosophy of science that again brings investigation of the natural world into the domain of a pluralistic reason, where various cognitive faculties receive their just deserts.

I am building on Dewey's moral philosophy, according to which all sciences "are a part of disciplined moral knowledge so far as they enable us to understand the conditions and agencies through which man lives. . . . Moral science is not something with a separate province, for physical, biological and historic knowledge must be placed in a human context where it will illuminate and guide the activities of men" (Dewey 1922, 296). Thus for Dewey, no firm demarcation between moral judgments and other kinds were possible, for "every and any act is within the scope of morals, being a candidate for possible judgment with respect to its better-or-worse quality" (279). He widened the scope of "morals" to value judgments writ large: "morals has to do with all activity into which alternative possibilities enter. For wherever they enter a difference between better and worse arises" (278).[8]

I have extended this moral-epistemological approach in previous studies (Tauber 2001; 2005a). In characterizing science in its broadest context, "moral-epistemology" highlights knowledge as structured by, defined through, and embedded in *diverse* values, and more to the point, these values include those established by lived experience and ordered by a personal ethic. Note that this terminology is not a characteristic usage, which addresses the epistemic status and relations of moral judgments and principles, that is, justification of statements or beliefs, in epistemology, or validation of judgments of actions, in ethics. Instead, "moral" stands here for acknowledging the degree to which knowledge is value-laden. Thus moral-epistemology captures the collapse of a dichotomous fact/value epistemology and substitutes an enveloping formulation: when science is configured within a moral-epistemology, the technical mastery of nature is coupled to the humane project of finding meaning in that knowledge. As we have seen with the example of environmental ethics (chapter 5), the science reaches deeply into the moral domain to find its respective *epistemological* footing, and beyond that domain a pervasive *ethic* guides and shapes practice and theory.[9]

If the line demarcating the moral from the epistemological seems faint, the reader must also be warned that the lines separating metaphysics from these other concerns is similarly obscure, for not until one journeys well within each territory does the landscape more clearly reveal its terrain. Perhaps that is the most disquieting message conveyed by this perspective on science: science cannot rest solely within epistemological demarcations; beyond its moral structure and interconnections into the social domain, its epistemology drives our metaphysics. A portrait of science must account for this full philosophical panoply. Despite my own ambitions to address this goal, the sketch offered here comprises only a draft of such a fuller description. Nevertheless, let us place a Dewey-inspired understanding of science in the integrative project we are considering. Three characteristics frame this approach.

First, following Dewey and later pragmatists, the only reality is "ordinary reality" (Shook 2000). This attitude seeks to envelop all experience, from esoteric frontline science to the full personal appreciation of a world so depicted. This final distillation of thought draws from diverse cognitive and emotional faculties and requires harnessing them into some integrated coherence. In short, an *integrative description* of the world and one's experience of it defines science in its broadest humane context.

Second, science presents the world as an ongoing *dialectical exploration*, in which meaning becomes the focus of interest, not truth (Dewey

1931, 4). This may appear to be a radical assertion, but in this discussion, truth not only has an epistemological standing, it also possesses an ethical one. And in this moral domain, truth is in service to meaning. This claim does not demote the standing of truth in any sense, but in this configuration, truth becomes a tool for making meaning. Truth claims then constitute a stage on the way towards some meaningful synthesis. On this view, truth functions in the service of meaning-seeking behaviors, which, of course, coincides with the integrative requirements of thought. Reality is thus experienced in an ongoing test of personal knowledge against the world that demands responses that invoke one kind of reason or another.

Meaning becomes the end point of knowledge as the individual interprets the world as a product of the present moment within the context of prior experience. In this fashion, meaning becomes the cognitive glue in which experience coheres. This attitude, developed by pragmatists and contemporary cognitive scientists, serves to capture human intention, because without the search for meaning, the motivation for integration and coherence would have no basis and experience would have no structure. Accordingly, a "foundational" epistemology has been replaced with a functional one, by which I mean the pragmatics of situating meaning-seeking ventures of ordinary experience (Tauber 2005a, 235–37).

Finally, this integrative effort requires a *self-reflective attitude* about science and how it becomes constitutive to our view of the world and of ourselves. Reflexivity than becomes the heart of the project, where comprehension of an integrated world, created by the search for meaning in a dialectical exchange with the world, emerges as epistemology's object of inquiry. On this view, meaning serves to focus judgment's function, an arbitration of experience to create human reality. The play of facts and values, interacting with varying valences assigned to each, serves as the métier of life experience, ordinary and otherwise. Following flexible (if not poorly defined) rules of navigation, this conjoined moral-epistemology acknowledges that reality ultimately places humans within the reality *they* experience. The facts of the world—or more simply, the world—only become factual with the values assigned by human evaluation. And thus reality becomes a product of mind and nature, as Kant first proposed—not constructed by a universal reason as he thought, but composed with varying rules, historically and culturally developed and thus contingent to time and place. Some cognitive rules may present themselves precisely, others less so, and some remain seemingly nonspecific, their full character shrouded as the faculties of human explanation and understanding weave the threads of experience into whole cloth, all the while oblivious to analytic attempts to discern them. Giving up foundations may well require abandoning rigorous epistemological formalisms as well.

The world so construed is fundamentally *moral* in the sense of human-valued, human-centered, human-derived, human-constructed, and human-intended. To segregate the self from the world as some separate entity defrauds philosophy's own quest, for a world without *human* value has lost human significance. The self firmly placed *in* the world constitutes the reality in which humans live. To fracture that fundamental unity not only distorts our understanding of ordinary experience, it misconceives it. So in the end, some form of a moral-epistemology offers an integrative description, urges a dialectical exploration, and provides a self-reflexive attitude to *know nature as meaningful experience.*[10] This effort to creatively establish coherence in a world refracted in many ways was introduced at the beginning of this book in the context of the conflict of science and religion. Now I trust it is apparent that I regard the challenge more generally as a moral venture, moral in the sense of defining and then affirming the moral compass, which guides our lives. With this orientation, science is thereby expanded into its full dimension. An epistemology which does not account for human value is laden with an unresolved irony, partly attributed to its historical development and partially due to its embedded precepts: That which masters also enslaves; that which provides also expends; that which enlightens also disenchants.

The philosophy I have described only outlines an ambitious project, but its theme appears clearly: the world picture science offers does *not necessarily* leave humans alienated. After all, science is a human invention for human use. The fruits of scientific labor may be harvested in many ways, and while the Faustian bargain seems firmly entrenched, critical attempts to revise that contract remain possible, and indeed necessary. The necessity arises from the metaphysical schism of the self peering *at* the world, where wholeness, integration, and coherence become aspirations. So to address both the instrumental and metaphysical roles of the scientific endeavor requires a reconfigured relationship of the knowing agent and her object of inquiry that acknowledges the complex interplay of objective knowledge and subjective meaning. From this point of view, moral-epistemology offers a response to Dewey's acute diagnosis:

> The problem of restoring integration and cooperation between man's beliefs about the world in which he lives and his beliefs about the values and purposes that should direct his conduct is the deepest problem of modern life. It is the problem of any philosophy that is not isolated from that life. (Dewey 1929/1984, 204)

On this view, instead of exclusively relegating philosophy of science to technical concerns, albeit highly useful for particular purposes, Dewey remained among those (e.g., Heidegger, Husserl, Jean Paul Sartre, Herbert Marcuse, Jürgen Habermas) who regarded positivism as a symptom of a malady, not of itself, but as an epiphenomenon of a deeper metaphysical crisis. Dewey's approach serves as a platform to continue philosophy's unfinished business of confronting the Cartesian embarrassment of self-consciousness, as a moral-epistemology's tools for insight and integration directly confront the dilemmas provoked by scientific knowledge.

Dewey's pragmatism makes its commitments to casting science in its full philosophical rendering. Instead of seeking some totalizing reason, he championed reason in all of its variegated forms. He regarded science as a philosophy that must find its place in its larger philosophical community by better integrating its offerings. This pragmatic approach recast Thoreau's spiritual and aesthetically inspired philosophy, but each pointed to the same problem and sought a similar resolution: an integrated and coherent placement of science in our lives.[11] The self-consciousness that constantly reminds humans of their individuality cannot be dissolved; no attempt at some mystical union is suggested here. Instead, both Dewey and Thoreau advocated an appreciation of the rich universe science offers us in all of its wonder and beauty; each sought an enrichment of life through better understanding of how the picture of the world has been enhanced by scientific advance. Beyond the technical mastery of nature, science then becomes a crucial tool for a fuller appreciation of nature and our placement in it.

We now live at the end of positivism's demise, so it is time to move on and ask, whither science's new philosophy? What epistemological precepts take the place of discarded dictums? If pragmatism has become our defaulted resting place, what are its own philosophical limits? After all, the pragmatic resides in an ether of values, and the incredulous may well ask to what end does the pragmatist venture? Dewey left no prescriptions, but only offered an approach, which outlines this project, or for the more skeptical, "attitude." After all, when surveying science studies over the past forty years, the pragmatic approach has overtaken other philosophies of science. By observing scientists in practice and the manner in which research becomes knowledge, science studies have demonstrated how a flexible value structure allows for the free development of investigation and theory construction. In that effort, it becomes apparent that the lines demarcating science from other forms of knowledge formation have been blurred. It is this crack in the fortress of the laboratory that reveals how science is embedded in its supporting culture and coheres in a vastly complex network of different forms of knowledge. So from this vantage, with the laboratory's door swung wide open, we have the outline of an integrative venture, based not on an idealist construction or a formal reduction but on the open architecture of a pragmatism that accepts the plurality of knowledge accompanied by multifarious values and diverse needs. In the spirit of this synthetic aspiration, I say to science, "Step forward and welcome back!"

Notes

Introduction

1 Epistemic values "promote the truth-like character of science" (McMullin 1983, 18) and are crucial for assessment. These values include predictive accuracy, internal coherence, external consistency, unifying power, fertility, and simplicity. Collectively, epistemic values are the means by which scientists assess the "fitness" of their theories to the natural world. Note that while such values have proven pragmatic worth, they also slide on an aesthetic continuum, where "simple" and "parsimonious," for instance, may carry different weights at different stages of a theory's development. An excellent example is offered in modeling complex ecological systems (Mikkelson 2001): when data are limited, simple models carry more verisimilitude, that is, predictive success, but when more data become available, the advantage of simplicity decreases, and more complex models dominate. (In ecology, holistic models are usually simpler than reductionist ones, thus when data are limited, holistic models have an advantage over models based on more reductionist methodologies.)

Nonepistemic values are those default values employed when epistemic criteria fail; more formally, they appear to close the empirical "gap between underdetermined theory and the evidence brought in its support" (McMullin 1983, 19). Although such values do not enhance a theory's epistemic status, they do reflect cultural, political, and religious beliefs, which combine to support or weaken truth claims. Successful theories witness a shift of support from nonepistemic to epistemic values (for an example, see Ruse 1999,143ff.), while other weaker theories may linger because of the power of the supporting cultural values. Appreciating how values evolved and were variously applied seemed obvious when considered in the nonepistemic sphere (cultural meanings, personal proclivities, political ideologies, and so forth). But similar dynamics of change and application were not widely accepted as governing the values operating in the laboratory. Science studies in the past forty years radically altered that basic formulation.

Underlying the epistemic/nonepistemic dyad is another controversy concerning the character of epistemic values themselves. In the realism/antirealism debate (see chap. 3. n. 11), epistemic values assume different roles and meanings (e.g., Cushing, Delaney, and Gutting 1984; Leplin 1984; Moser 1990; Boyd, Gasper, and Trout 1991; Boyd 2002), and from those differing meanings the battle lines were drawn over "constructivism," more specifically, "social constructivism."

2 Perhaps the key urtext of science studies is Steven Shapin and Simon Schaffer's *Leviathan and the Air Pump: Hobbes, Boyle and the Experimental Life* (1985). This seminal work attempts to dispel what the social constructivists maintain is a false division between epistemology and sociology. The authors argue that beyond showing how the social context affected the scientific debate between Boyle and Hobbes in seventeenth century England, what was really at stake was the very invention of a science, its social context, and the demarcation between the two. Bruno Latour lauds the book as "the real beginning of a comparative anthropology that takes science seriously" (1993, 15). For an excellent review of the relation of science, values, and ideology in the contemporary context, see Helen Longino's "Science and Ideology" (chapter 9 of *Science as Social Knowledge*, 1990), which is based on the work of Habermas, Michel Foucault, and the feminists Evelyn Fox Keller and Donna Haraway. Opponents argue that such meta-analyses, in general, cannot achieve objective priority. They call into question whether a neutral constructivist project is even possible at all. Other criticism appears to be leveled against particular ideological agendas. Thus controversy concerning the degree to which science may be criticized from a feminist or Marxist perspective revolves around the explicit or implicit charges concerning the prejudice (i.e., male-dominated or capitalist-driven science) that might heavily determine, if not direct, the *conceptualization* project of objective science itself.

In my study, I refer repeatedly to "science studies," which must be distinguished from "science and technology studies" (STS). Science studies refer to the broad inter-disciplinary examination of science, although some would place it more specifically as a branch of the sociology of knowledge. The character of study depends on the disciplinary commitments of the commentator, so for me the most interesting (and challenging) aspects of the examination of science gives primacy to philosophical insight as it impacts on interpretations of the history and the sociology of science. Note, the prominence of the philosophical orientation does not deny that philosophers are heavily influenced by sociological portrayals of how scientists actually function. Accordingly, each approach (philosophical, historical, and sociological) contributes to the others. While I regard science studies serving as a catch-all term for the various academic disciplines that study the production and use of science knowledge, STS treats science as a social institution and specifically examines (1) how social, political, and cultural values affect scientific research and technological innovation, and (2) how these in turn affect society, politics, and culture. STS is thus decidedly sociological in its orientation, and as a more circumscribed discipline, it has its own professional societies, journals, and academic departments, which confer both baccalaureate and graduate degrees. Here, chapters 4 and 5 primarily deal with issues central to STS, while the remaining exposition adheres to the philosophical approach as described above. Following this organization, one thesis presented here is that the STS understanding of science draws heavily from the philosophical framework in which constructivism plays such a prominent role. For representative anthologies that cover the broad discipline of science studies see Pickering (1992), Jasanoff et al. (1995), and Biagioli (1999). For circumspective views of the accomplishments of science studies, see recent interpretative studies by Brown (2001), Zammito (2004), and Sokal (2008).

3 The "unification of reason" problem formally dates to the eighteenth-century as posed by Kant. He characterized science in terms of describing the conditions by which "pure reason" functioned to study the natural world. In contrast, he presented a second kind of rationality, "practical reason," to navigate the social/moral world. So, for Kant, two kinds of reason were required for human understanding of the natural and moral universes. The problem, then, was to integrate these contrasting ways of knowing. The challenge of how reason might be regarded as unified, the "unity of reason" problem, does not first appear with Kant's schema, but grows from modernity's conundrum of determining how humans can be both part of the natural world of cause and effect, and at the same time, exercise free will and thus assume moral responsibility. How Kant regarded "pure" and "practical" reason as unified has been deliberated in three basic formulations:

> (1) They are *compatible* with each other, that is, insofar as the principles of one do not conflict with those of the other;
> (2) both can be derived as components of a unitary and complete system of philosophy, which has as its starting point a single first principle;
> (3) they possess an identical underlying "structure," or constitute what is in essence a single activity of the subject (Neuhouser 1990, 12).

The autonomy of both theoretical and practical reason serves as the bedrock of Kant's entire philosophy, a system that provides for freedom in both the apprehension of the natural world and the discernment of moral action in the social world. Our task remains the same, namely, the challenge of making reason—again, in Kant's words, pure reason and practical reason—whole (Neiman 1994; Tauber 2007). I make no pretense of offering some comprehensive solution, but I do discern an approach.

4 I have adopted a discursive strategy, which cuts across several disciplinary lines. Written in an Anglo-American voice, dominated by a pragmatist point of view, "continental" themes in the philosophy of science appear here in obvious, as well as disguised, formats (Gutting 2005). Recently, others have also found common cause to create a new trans-Atlantic treaty as the problems identified in one sector resonate with concerns articulated in the other (e.g., Apel 1988; Friedman 2000; Mulhall 2003; Prado 2003; Rudd 2003; Egginton and Sandbothe 2004; Rockmore 2005; Braver 2007). From my point of view, this movement can only enrich the pursuits of all. Indeed, we may well have arrived at an inflection point in philosophy's history, where the old divisions are melding into something new. I believe that expanding the characterization of science along such lines offers a promising path towards that goal.

Chapter 1

1 Cardinal Schönborn's reference to intelligent design in his *New York Times* op-ed piece (July 7, 2005) may be summarized as a debate about the evolution of biological complexity, which has a long history (Ruse 1996) and more recent literature on the issue of intelligent design (Pennock 2001; Ruse 2003; Dembski and Ruse 2004; for a succinct review, see Nakhnikian 2004). Evolutionists argue that in the course of random mutations, more complex options are offered and these may be chosen to accommodate the stresses of changing environments and competition among other species. On this view, biological diversity, initiating sometimes more complex, and at other times more simple "solutions," have appeared. According to neo-Darwinism, "design" is an unnecessary element in explaining evolution (see for example, Miller 2004).

2 The creationists pose a somewhat different kind of argument than intelligent design proponents. Their argument does not acquiesce to the scientific findings, but rather disputes the facts themselves. They have stubbornly opposed contemporary Darwinism by insisting that creationism is a bona fide theory of life and that the findings documented by evolutionists assume a different meaning in creationist theory. Students of this controversy have concluded, and I think fairly, that the argument cannot be won by evidence (Sober 1993). The Darwinists point to mountains of molecular, paleontologic, and phylogenetic data to show blind evolution at work in the field and laboratory, as well as in the geological record. The creationists argue that God placed the history there by reason of divine wisdom; that evolution is directed and thus bestowed by God; that an omniscient being created the world, and perhaps this Creator continues to guide evolution, for its own purposes. Given the fundamentally different underlying presuppositions of each point of view, no meaningful debate exists.

3 Science's instrumentality has at least two dimensions. The first refers to how research is applied to devise technologies. These technologies might be put to constructive use (the usual case) or instead employed as a tool for purposes quite at odds with the original intent of seeking knowledge for the social good. This instrumental quality of science (its technological power) holds one of its ironies: instead of maintaining its original philosophical credentials, science, or more precisely, its technological progeny, too often is so divorced from those earlier concerns that the basic research becomes a tool that may be applied independently of the primary intent of the investigation. Co-opted by those whose own agenda has nothing to do with promoting the Western values that spawned science in the first place, we have painfully learned how some may use powerful technologies as an instrument of power for socio-political ends at odds with our own.

A second sense of instrumentality refers to science's intellectual activity, a mode of discovery and knowing, where the findings are used like a currency to buy different goods. The goods are findings or ideas, which are then placed into a conceptual context. The competing context may be differing scientific theories, but in this discussion, I am interested in religious contexts. For example, the sun assumes a certain meaning as conceptualized by a materialistic astrophysics, and a different one when regarded as Apollo racing across the heavens. The Greek myth has been eclipsed, but certain religious fundamentalists will contest the ontology of particle physics as "the cause" of the sun's birth, and instead refer to God's will and deliberate choice. Where the physicist will admit that knowledge reaches a limit, the true believer will push the universe's origins back into the divine act. The question at hand thus may be simply defined: where does knowledge end and belief begin? That border has again become an active battlefield, for no less than the authority of knowledge is at stake.

4 Sublimity became an alternative conception of aesthetic response and interpretation in the eighteenth century (Ashfield and de Bolla 1996) and thereby began the most recent turn toward nature as a religious resource and object of divinity's presence during the nineteenth century (McFarland 1969). With Kant and Hegel, sublime reflection provided a model for interpretative responses to the divine Other, and the uncanny or ungraspable dimension of intuitive experience (Pillow 2000). Sublimity achieved a particularly enhanced status during the romantic period and has resurged in some quarters of environmentalism (Tauber 2003). That history is not our present concern (although environmentalism is examined in chapter 5). I have a more modest goal, namely to highlight how aesthetic and religious wonder may be regarded analogously, at least in certain classical formulations, and how science may enlist in that coupling.

5 For Nietzsche, the aesthetic domain expressed the core of human vitality and the means of release from a restrictive rationality. Heidegger eloquently described the role of beauty for Nietzsche's philosophy:

> Rapture as a state of feeling explodes the very subjectivity of the subject. By having a feeling for beauty the subject has already come out of himself; he is no longer subjective, no longer a subject. On the other side, beauty is not something at hand like an object of sheer representation. As an attuning, it thoroughly determines that state of man. Beauty breaks through the confinement of the "object" placed at a distance, standing on its own, and brings it into essential and original correlation to the "subject." Beauty is no longer objective, no longer an object. The aesthetic state is neither subjective nor objective. Both basic words of Nietzsche's aesthetics, rapture and beauty, designate with an identical breadth the entire aesthetic state, what is opened up in it and what pervades it. (1979, 123)

6 Another way of contemplating the separation of the *cogito* from the world it surveys turns self-consciousness on its head to make partition a virtue. Hans Jonas observed how consciousness provides the guarantee of personhood. In this view, the very *inability* to merge with Nature assures humans a transcendence as moral creatures and becomes the metaphysical basis of selfhood (Jonas 1963, 320ff.), or as Olaf Hansen observes, "The clear view of the unattainable lends identity to our existence in this world, *because* we cannot integrate the cosmos. So then, whatever shape each individual's existence will have, its identity is derivative of a purity of vision, which can only be defined in terms of its unworldliness. Hence, the worldly, practical consequences of our quest for identity" (Hansen 1990, 4).

7 Daniel Greenberg offers a typical opinion about science's ethics coupled to a motive:

> Failings in openness, collegiality, respect for human and animal experimental subjects, and scientific and financial integrity are common topics in scientific journals and on a thriving conference circuit. The hand-wringing, arguments, and recriminations go on, within and beyond the boundaries of science. But ethical concerns are a sideshow of science, providing grist for the press and moralizing politicians, though with little actual effect on the conduct of the research enterprise. Within science, the ethical issues are overshadowed by material concerns. These concerns consume more energy than any attempts to rectify ethical shortcomings. More money for science is the commanding passion of the politics of science. . . . Despite its ethical angst and shortcomings, science today is only a short way down the path to becoming a toady of corporate power. Disturbing evidence is plentiful for concern about the ethical costs of science's single-minded pursuit of money. (2003, 3)

8 As a general matter, the relationship of facts and values as a philosophical and psychological *problem* is as old as history. After all, Thucydides, in describing the collapse of civil order during the Peloponnesian War (3.82.4), observed, "And people exchanged the conventional value of words in relation to facts, according to their own perception of what was justified" (translated by Price 2001, 9–10). Thucydides understood that the moral framework fixed the meaning of facts, and his politico-social observation easily translates in our own age to a parallel understanding about science.

9 The fact/value dichotomy issue now possesses an impressive literature, and consequently any attempt to summarize it commits a certain injustice. Suffice it to note here that so-called value-free science adopts three basic claims concerning the construction and use of facts (Kincaid et al. 2007, 13): Objective science never presupposes nonepistemic values (1) in determining what the evidence is or how strong it is, (2) in providing and assessing the epistemic status of explanation, nor (3) in determining the problems scientists address. Each of those assertions, over a wide spectrum of arguments, has been challenged (Putnam 1982, 1990, 2002; McMullin 1983; Longino 1990, 2001; Proctor 1991; Lacey 1999; Farrell 2003; Kincaid, Dupré, and Wylie 2007).

Chapter 2

1 The so-called "new physics" emerges from the probabilistic character of quantum mechanics and the Heisenberg "uncertainty principle," which states, basically, that observers cannot determine in any absolute sense the place and character of an atomic particle at the same instant: "if I know where an electron is I have no idea what it is doing and, conversely, if I know what it is doing I do not where it is" (Polkinghorne 1985, 3). Simply, physical observation precludes the simultaneous description of the position and momentum of an elementary particle and reality as described by this new physics reveals a universe determined, at least in terms of that reality as *known*, by the observer observing (Jammer 1974, 56ff; Davies and Brown 1986, 1–39).

Because of the "measurement effect" and the consequent unpredictability of describing quantum mechanical events, physics would abandon its earlier aspiration of modeling an observer-independent reality. The entire basis of "factual" knowledge now required reassessment, and as a consequence the very status of the idea of "objectivity" came into question. Two views, or interpretations, divide commentators (Shimony 1991; Fine 1991):

> The "antirealist" school, most radically formulated by Bohr (and less so by various supporters, e.g. Werner Heisenberg, John von Neumann, Eugene Wigner) holds that the entire formalism [quantum mechanics] is to be regarded as a tool for deriving predictions, of definite or statistical character, as regards information obtainable under experimental conditions described in classical terms. . . . There is no quantum world. There is only an abstract quantum physical description. It is wrong to think that the task of physics is to find out how nature is. Physics concerns what we can say about nature. (quoted by Polkinghorne 1985, 79)

In that spirit, Werner Heisenberg advocated a strong antirealism, where experiments of atomic phenomena are as real as any phenomena of daily life, but the atoms or elementary particles are not, inasmuch as they form a world of potentialities or possibilities rather than one of things or facts (Heisenberg 1958a, 1958b). "Potentiality" becomes an "actuality" once measurement is made, i.e., observations collapse the wave functions into one state or another, and the "reality" of the observed is finally determined by the measurement. Putting aside whether this process *requires* the intervention of mentality (Shimony 1991, 526–27), the basic notion of mind and nature making a composite of reality has taken on a new dimension.

Albert Einstein strenuously resisted Niels Bohr's interpretation and maintained that the probabilistic character of quantum mechanics arose from its incompleteness, i.e., its theoretical incompleteness, which "is not just a desideratum for the purpose of avoiding strange, or unintuitive . . . implications, but that it is actually required by the formalism of quantum mechanics itself" (Shimony 1991, 521). In the famous Einstein-Podolsky-Rosen experiment (Davies and Brown 1986), Einstein attempted to refute Bohr's antirealism by suggesting that "hidden variables" were present, which would account for quantum uncertainties and action at a distance. That experiment gave rise to a vast literature. At this point the option of undeclared variables has been excluded and other solutions for a "realist" picture have been offered, but the debate remains unresolved among those who seek to support or dethrone realist interpretations (see chap. 3, n. 11).

2 The conceits of an independent observation and assessment have been radically challenged by twentieth-century sociologists of knowledge, who maintain that knowing and ways of knowing are fundamentally communal and that the vector private → community (the positivist conceit) is balanced, and some would argue dominated, by the reverse vector, community → private. The dominance of the "thought collective" (Fleck 1935/1979) or the social solvent of "power" (Foucault 1994) places the individual so embedded within the collective that he or she is not only defined by the social, but only exists and knows as an element therein. This formulation thus has radically challenged nineteenth-century conceptions of the

relationship of the individual and his society, whereby the romantic individualist and the positivist share the notion of independence and individuality.

3 Hume's Law appears variously in Hume's opus, but most clearly stated in *A Treatise of Human Nature* (Book III, Part 1, section I): "I have always remark'd, that the author proceeds for some time in the ordinary way of reasoning, and establishes the being of God, or makes observations concerning human affairs; when of a sudden I am surpriz'd to find, that instead of the usual copulations of propositions, *is*, and *is not*, I meet with no proposition that is not connected with an *ought*, or an *ought not*. This change is imperceptible; but is, however, of the last consequence. For as this *ought*, or *ought not*, expresses some new relation or affirmation, 'tis necessary that it shou'd be observ'd and explain'd; and at the same time that a reason should be given, for what seems altogether inconceivable, how this new relation can be a deduction from others, which are entirely different from it. But as authors do not commonly use this precaution, I shall presume to recommend it to the readers; and am persuaded, that this small attention wou'd subvert all the vulgar systems of morality, and let us see, that the distinction of vice and virtue is not founded merely on the relations of objects, nor is perceiv'd by reason." (Hume 1739/1978, 469–70)

4 Whether these neutralizing rhetorical devices largely reflect a commitment to a mechanical (Cartesian) view of the knowing self that strives toward an unrealizable mirroring of nature, or represent a dynamic and evolving science based on some universal objective ideals has been grounds for rigorous debate (e.g., Sellars 1963; Trigg 1980; Gutting 1983; Boyd 1984; Wolterstorff 1984; Rescher 1987; Laudan 1990a). For example, Kenneth Gergen (1994) makes his case based on the structure of the knowing self as dualistic, where an inner psychological state of "knowing" is contrasted to an external world to be "known." Positivists viewed the objective mind as seeing "things for what they are" and is thus "in touch with reality," and there is a correspondence of this reality with our language that depicts it. Gergen rejects this position, and argues that objectivity rests on an unsteady psychological foundation. Gergen's major attention is on how one determines the accuracy of one's internal identifications, for it is in the collective agreement concerning perception that the limits of individual knowledge are subsumed to consensus.

5 In a seminal social history of the scientific community of the seventeenth century, Shapin and Schaffer emphasized how the civility of English aristocracy framed the powerful legitimatization process that social authority conferred on scientific discourse. Thus facts were in large measure accepted by the scientific community based on the social standing of the observer (Shapin and Schaffer 1985; Shapin 1994; see Introduction, n. 2). This implicit trust, in our own era, has been transferred to the community at large. As Shapin observes, "the very power of science to hold knowledge as collective property *and* focus doubt on bits of currently accepted knowledge is founded upon a degree and a quality of trust which are arguably unparalleled elsewhere in our culture" (1994, 417). Trust amongst scientists is of course based on the shared ideal of scientific objectivity. Because the individual cannot achieve objectivity as a private mental condition, monitoring objectivity then becomes a matter of broad social policy, so consensus is required among participants as to what is considered an objective account. After all, the criteria of objectivity have varied

over the past four centuries as the standards of scientific practice have continued to evolve and, of course, different scientific disciplines employ different standards of proof (Megill 1994).

6 What constitutes a fact depends upon the metaphysics, epistemology, theory of truth, and semantics that determine its definition. Insofar as facts are what make true statements true, inquiries that do not distinguish the metaphysical nature of truth from epistemological concerns or from the linguistic use of truth will hopelessly muddle important distinctions. Notwithstanding debates about correspondence theories of truth (for example, Russell, Austin, and Strawson) and the empirical status of facts (scientific realism versus social constructivism), it is apparent that the various sciences encompass a wide variety of facts, whose epistemological status and linguistic uses vary, perhaps best illustrated by the history of fact (Poovey 1998). (For discussion of truth, see chap. 3, n. 13).

7 The prevailing project of twentieth century existential phenomenologists described how authenticity demanded a new self-consciousness. The drive toward objective contemplation, logical analysis, scientific classification putatively cuts humans off from intimacy with the world. The dissection of the world yields a kind of knowledge, which must still be integrated *meaningfully*. The scientific object may reside seemingly separate—"out there"—but the issue is to integrate that object to human experience, rational *and* emotional. The search for this common ground is the elusive synthesis of our very selves in a world ever more objecified *from* us. More fully discussed in the conclusion, where Husserl's position is explored (Harvey 1989).

8 In contrast to later scientific reports restricted to positivist descriptions of observations, Goethe's *Theory of Colors, Farbenlehre* (Goethe 1810/1988) combines his views of the history of the problem and its factual findings, aesthetics, psychological perception, epistemology, and mathematics (Fink 1991). Thus this work represents a holistic evaluation of color, merging what later became different fields of investigation with an aesthetic appreciation that Goethe forthrightly used to make his scientific judgments and help him derive his theory.

Goethe was heavily influenced by Kant, whose *Critique of Judgment* (1790/1987), he "returned to again and again" (Goethe 1817a/1988, 29). Rather than divide reason as irredeemably separate, Goethe regarded them as emanating from the same root, so that poetry and science "both are subject to the same faculty of judgment" (idem). He closely followed Kant's own lead, who viewed aesthetic judgment as unlike either scientific or practical judgment, in that it functions entirely subjectively, that is, solely in reference to the knowing subject. Kant avoided solipsism, because, according to him, this judgment commands assent through the common ground of communal or shared subjectivity (and thereby mediates some degree of shared understanding). Goethe reiterated this position in many forms as it served as a guide for his research (see chap. 1 [Goethe 1790/1989, 1794/1988, 1817b/1989]). Though his methods are now suspect, a deeper philosophical appreciation rests on understanding the constructivist character of his science. This view would gain prominence after the publication of Kuhn's *Structure of Scientific Revolutions* (1962). See chapter 3 for a discussion of Kuhn's work.

9 If the natural world is seamless, then presumably the approach to its study should also be unified—both epistemologically and metaphysically. Part of the current

assessment of the positivist agenda concerns the growing opinion that the various scientific approaches applied by the different species of scientific inquiry are not easily linked to each other to offer a coherent and seamlessly unified "picture" of the world (Dupré 1993; Galison and Stump 1996; Schaffner 2002). This issue is also considered in the conclusion.

10 The philosophical literature on the definition of value is extensive and contested (e.g., Brandt 1979; Sadler 1997; Stempsey 1999; Putnam 1985; Sayre-McCord 1988; Railton 2003). Values come into play at both the level of scientific knowledge, that is, in assessing what constitutes a scientific explanation and scientific fact, and at the level of scientific application. I follow Putnam's summation:

> [I]f values seem a bit suspect from a narrowly scientific point of view, they have, at the very least, a lot of "companions in the guilt": justification, coherence, simplicity, reference, truth, and so on, all exhibit the *same* problems that goodness and kindness do, from an epistemological point of view. None of them is reducible to physical notions; none of them is governed by syntactically precise rules. Rather than give up all of them . . . and rather than do what we are doing, which is to reject some—the ones which do not fit in with a narrow instrumentalist conception of rationality which itself lacks all intellectual justification—we should recognize that *all* values, including the cognitive ones, derive their authority from our idea of human flourishing and our idea of reason. These two ideas are interconnected: our image of an ideal theoretical intelligence is simply part of our ideal of total human flourishing, and makes no sense wrenched out of the total ideal, as Plato and Aristotle saw. (1990, 141; emphasis in original)

11 For the logical positivists, who dominated philosophy of science from the 1930s through the 1950s, philosophy became a branch of theoretical science itself by demarcating discovery from verification. Accordingly, scientists used empirical methods for discovery, leaving the process of verification to philosophical analysis. Reichenbach (1938) argued that logical positivism's task was to clarify and analyze scientific claims or concepts *as they were found in science.* Thus the analytic project restricted itself to language, while scientists explored the world to have their truth claims analyzed in their own context. The difference of the two projects was "as basic and irreducible as the difference between language and the world that language describes" (Romanos 1983, 4). A further assignment included another demarcation, that of science and pseudo-science, for example, astrology or phrenology. The authority of science, or more particularly, the authority of a certain kind of knowledge employing strict criteria as outlined above, was to identify pretenders who sought to append scientific authority to their own agenda.

12 Thoreau's specimen collecting and classifying assuredly qualified as scientific. Indeed, as he matured, his projects became more ambitious and comprehensive. For example, Thoreau carefully studied the dispersion of seeds in an attempt to show that the generation of plants was dependent on seeds alone, and that the variety of mechanisms available for propagation required scrupulous examination of plant patterns, weather conditions, topographical opportunities, and potential ani-

mal and insect carriers. These were just a few of the factors he considered (Thoreau 1993). A second example, the "Kalendar" project, epitomized Thoreau's formalized attempt to document variegated nature and detect regularity amidst natural change. He sought to parse the seasons by culling his journal to create a series of monthly charts. In these charts he listed various natural phenomena in a left-hand column and strung the years along the top. The phenomena he tracked included the height of the Concord River, rain patterns, rainbow appearances, temperature, leafing of trees, and so on. Some of such note keeping made its way into his published writings; for instance in *Walden*, Thoreau lists the dates the pond was freed from ice for the years 1845, 1846, 1847, 1851, 1852, 1853, and 1854 (1971, 303). However, no matter how well motivated such record keeping might have been, Thoreau himself thought this chronicle proved to be an essentially futile endeavor that, despite his most earnest efforts, remained partial and incomplete. The constancy or regularity he sought could not be demonstrated, but today, Thoreau's misgivings notwithstanding, climatologists are finding the data of extraordinary value (Miller-Rushing and Primack 2008). Thoreau's studies were thorough, and he drew his conclusions scrupulously to meet the methodological standards of the day. Beyond the love of field study, Thoreau aspired to making novel discoveries and suffered disappointment when he failed (Tauber 2001, 143–45).

13 I find Proctor's last phrase, "The real is not the rational" (1991, 135), intriguing as a play off Hegel's idealism. In Hegel's Preface to *Philosophy of Right* (1821) he famously writes, "What is real is rational; and what is rational is real" (translated by S. W. Dyde 1896, http://www.marxists.org/reference/archive/hegel/index.htm. Newer translations render the German as "What is rational is actual; and what is actual is rational" (Hegel 1821b/1952, 10; 1821c/1991, 20). The editors of each translation offer two explanations. First, as T. M. Knox writes:

> Note Hegel is not saying that what exists or is "real" is rational. By "actuality" . . . he means the synthesis of essence and existence. If we say of a statesman who accomplishes nothing that he is not a "real" statesman, then we mean by "real" what Hegel calls "actual." The statesman exists as a man in office, but he lacks the essence constitutive of what statesmanship ought to be, say effectiveness. . . . Hegel's philosophy as a whole might be regarded as an attempt to justify the identification of rationality with actuality and vice versa. (Hegel 1821b/1952, 302, n. 27)

Second, as Allen W. Wood notes,

> In his later expositions of this famous (or infamous) saying, Hegel is at pains to point out that it does not mean that everything is as it ought to be, or (more particularly) that the existing political order is always rational. . . . Far from hallowing the status quo, Hegel's formulations of the rationality of the actual in his lectures of 1817–1820 emphasize the dynamic and progressive aspect of the reason which is at work actualizing itself in the world: "What is actual becomes rational, and the rational becomes actual." (Hegel 1821c/1991, 389–90, n. 22)

More than simply an indictment against idealism or some kind of social metaphysics, Proctor's statement, "The real is not the rational," highlights how *science* itself is not rational in two distinct ways. First, the emphasis of science as an empirical study of nature was an important distinction made in early modernity to distinguish science from the rationalism of scholasticism. The second way, and the one I wish to emphasize, is how science is not "rational" in human terms, that is, meaningful. In this wider context, to be rational is to conform to a theory of meaning or understanding, one imbued with various assumptions, which themselves are value-laden. To recognize science as self-consciously objective and neutral is to assert the deliberate corrective to subtle yet intractable human biases in all of their manifestations.

Chapter 3

1 The emerging discipline of "emotional intelligence" (EI) examines how emotions are assimilated in thought (Matthews, Zeidner, and Roberts 2002, xv). Although closely related to "social psychology," segregating this issue as a formal discipline only dates to the early 1990s, when both popular and scholarly interest channeled what had hitherto been an implicit understanding into a new scholarly consensus: the traditional putative antagonism between the faculties of "reason" and "emotion" requires a new synthesis of understanding their intimate relationship. Despite the long philosophical tradition that would separate them, evidence increasingly suggests that emotions and rational thinking are closely intertwined (de Sousa 1987; Damasio 1994; Panksepp 1998; Bar-On and Parker 2000; Nussbaum 2001; Matthews, Zeidner, and Roberts 2002). Like other forms of intelligence, such as mathematical, musical, spatial, analytical, and so forth (Gardner 1993), emotional intelligence offers intuitions that participate, and in certain instances dominate, various domains of human action (e.g., social interactions). In a general sense, emotions drive individual motivations and play an important role in adjudicating judgments of various kinds. For example, philosophers have long pondered the character of *intention* as the property of directing thought to achieving its ends. Intention is in part "rational," for example, How shall I most effectively address my hunger? But it is more. To address such a question requires a mixture of rational strategy and emotional assessment. For instance, while I might grab my colleague's lunch, my desire for collegiality and need for approval mitigate my immediate impulse. The ability to mediate one's intercourse in the social world requires a social sensibility, an awareness of how we are perceived by others, the effective interactions that allow us to proceed successfully to fulfill our goals, and the necessary requirements (e.g., motivation, concentration, limits of satisfaction, self-confidence, etc.,) that are the means to achieving those ends. Plainly, the rational and the emotional aspects cannot be rigidly divided.

2 Paul Thagard has provocatively shown the role of positive and negative emotions in James Watson's autobiographical account of the discovery of the structure of DNA (Watson, 1968; Thagard 2002). There, Watson documents how he felt at various points in his research, both in terms of the cognitive process (happy, disappointed, relieved, angry, etc.) and in his interactions with his competitors (Linus Pauling, Rosalind Franklin, and Maurice Wilkins). Most of the emotion is described in the context of discovery, from the choice of the topic to the euphoric insight. It seems clear that the creative process drew on many abilities, but the drive and competitive-

ness of these scientists is striking. However, while their psychological state commands biographical interest, the more pressing challenge is to better understand the cognitive role such emotions play in scientific creativity (Hookway 2002).

3 See chap. 2, n. 9.

4 Unbeknownst to Kuhn, the original version of his essay, *Structure of Scientific Revolutions* (1962), was well received by leading positivist Carnap, who had shown signs of shifting sympathetically towards certain elements of Quine's devastating critique of positivism's central tenets (Friedman 1999). Others have argued that Kuhn and Carnap have strong philosophical affinities (Irzik and Grünberg 1995) and a shared larger intellectual perspective (Galison 1990).

5 My favorite example of "perceptive blindness" is Leonardo da Vinci's anatomical drawings, where the muscles and skeleton are meticulously accurate, but the viscera, described by Galen and accepted by pre-Harvey anatomists, was reproduced by the artist as he was taught. An exhibition of these works made a lasting impression on my understanding of this principle (Leonardo 1987). Norwood Russell Hanson made similar arguments about theory-laden observations at about the same time as Kuhn, and despite the importance and influence of his work, he did not draw conclusions that would radically change the direction of science studies (Hanson 1958). Note, whereas Quine rejected the theory/observation distinction for *linguistic* reasons, Hanson based his argument on the nature of *perception*.

6 "Strong" constructivists (as already mentioned were disowned by Kuhn) "adopted a relativist epistemological position, which emphasizes the under-determination of solutions to scientific problems and de-emphasizes or altogether denies the importance of the empirical world as a constraint on the development of scientific knowledge. . . . [Further all constructivists argue that] the actual cognitive content of the natural sciences can only be understood as an outcome of social processes and as influenced by social variables" (Cole 1992, 35; see also Ashmore 1989 for a review). The position is well explained by Laudan, who argues that the contending epistemic (as opposed to ontological) positions, each fails to answer why science works so well. The realist fails to answer the question (he cannot explain to what approximation theories are true nor account for the dissociation between truth of theories and their epistemic success, or lack thereof), and the relativist does not even grant the legitimacy of the question. Laudan suggests that realists should focus on epistemic analysis of the methods of theory testing, and that the relativist must grapple with the obvious success of science to better control and predict nature than other systems of belief or exploration (1984b).

7 Kuhn's *Structure of Scientific Revolutions* (1962, 1970), is generally regarded as an imperfect work, having "a philosopher's sense of sociology, a historian's sense of philosophy, and a sociologist's sense of history" (Fuller 2001, 32), but it catalyzed an intellectual realignment that has still not found its resting place (Hoyningen-Huene 1993; Horwich 1993; Marcum 2005b; Davidson 2006). It accomplished that task because Kuhn "shamed post-war philosophers of science into dealing with *real science*, rather than trivial logical surrogates for real science" (Giere 1997, 497; emphasis in original).

8 Ludwig Wittgenstein lurks as the shadow figure in the history of logical positivism. He argued that only particular kinds of empirical questions were meaningful

and qualified as warranting *certain* answers. His last writings are rift with skepticism and guarded statements about scientific knowledge, the standing of beliefs, the individual status of certainty, the logical basis of empirical propositions, and so on (Wittgenstein 1969, especially 24–31). Indeed, when he wrote, "The difficulty is to realize the groundlessness of our believing" (24e), he reiterated the uncertainty of even empirical claims, and thereby straddled the fine line that separates solipsism and public discourse: "I act with complete certainty. But this certainty is my own" (25e). (No wonder later radical social constructivists would enlist him for their own cause [e.g., Bloor 1983].)

In *Tractatus Logico-Philosophicus* (1921/1961), Wittgenstein presented a "picture" theory of language that allowed for the legitimacy of certain propositions based on their facticity, and for the rest—ethics, metaphysics, aesthetics—he advised that we must remain "silent." Later, in *Philosophical Investigations* (1953/1968), Wittgenstein disallowed language to serve as a "representation" of the world, and thus he dispelled any notion that language serves as a direct correspondence to reality. Language (coincident with the mind) could not mirror nature. Further, he discounted our ability to discern any formal language rules for natural languages, because natural languages do not take form analogous to formal language. For Wittgenstein, actions or behavior defined language and its "logical" basis in the ordinary sense of daily communication. Throughout his writings, he drew distinctions between facts as derived from objective methods as opposed to other kinds of experience—emotional, supernatural, ethical—in which personal (and thus unverifiable) belief confers a radically different status to certainty. And because language and thought were inseparable for Wittgenstein, much of "thought" would remain inarticulate; further, the status of private thought was highly problematic.

While some would argue that Wittgenstein regarded philosophy as failing to find solutions to traditional metaphysical problems because these problems were poorly formed, it is more in keeping with his thought that he regarded such issues as "nonsense," by which he meant not realizable in any logical sense. He did not deny their presence as human challenges, but he rejected *philosophical* solutions to such questions. Accordingly, the narratives woven around the classic philosophical issues—ethics, aesthetic judgment, personal identity, and so on—are simply misconceived if we expect some kind of *logical* formulation. In the end, Wittgenstein sought to comprehend the locks and chains in which language ensnares human dialogue and the problems we pose for one another as products of false application of logic.

9 Quine's doctrine that "natural science is empirically underdetermined" (Quine 1975, 313) maintains that "even if we do *not* reject any observation sentences that we take to have been verified, and even if we do *not* allow any changes to the theory, still, if there is at least one theory that has a given set of observational consequences, then there will always be more than one [theory]" (Putnam forthcoming; emphasis in original). This *underdetermination doctrine* must be distinguished from the *Quine-Duhem thesis* which expands Duhem's argument that a "crucial" experiment could never refute a theory because adjustments could always be made regarding the background assumptions to excuse the failure of an experiment. The thesis arises from the dependence theories have on observations, other theories, or auxiliary

hypotheses. To test theory A, theory B must be used, and the expected observation is then a product of A and B together. If the observation conflicts with the expectation, then the cause may be that theory A is false, or B is false. Since the finding is dependent on A and B, the only conclusion that can be drawn is that A and B are not both true together. While the Quine-Duhem thesis is widely held, the underdetermination doctrine risks being either trivial or unprovable, as Quine himself admitted (Ben-Menahem 2006, 246–48).

The *indeterminancy of translation* builds directly from "inscrutability from reference," as Quine maintained that that there is no formula, no determinate method, no specified procedure (dictionary) to translate from one language to another. Instead of some kind of systematization of translation, Quine was satisfied with pragmatic communication achieved without rules. This was hardly a novel insight, in one sense, as the art of text translation testifies. But he went further than simple adequacy or veracity to propose that plausible translations may actually conflict when referenced to some metalanguage, and this point highlights his views on language more generally (1969c).

10 Quine's allusion to "Neurath's boat" is deeper than the simple use of an allegory. In 1931 Neurath had already presented, in the context of his Unified Science project, the holism that Quine would utilize, and in the process, Neurath distanced himself from Carnap and other Vienna Circle philosophers in ways not fully appreciated until Quine restated that position: "The study of language can perfectly well be combined with the study of physical processes; for one always stays in the same field. In staying within the closed area of language one can express everything. Thus *statements are always compared to statements*, certainly not with some 'reality,' nor with 'things,' as the Vienna Circle also thought up to now" (Neurath 1931b/1983, 53). Thus Neurath had already arrived at a holistic viewpoint well before Quine, but the full implications had not been drawn:

> It is always science as a system of statements which is at issue. *Statements are compared with statements*, not with "experiences," "the world," or anything else. All these meaningless duplications belong to a more or less refined metaphysics and are, for that reason, to be rejected. Each new statement is compared with the totality of existing statements previously coordinated. To say that a statement is correct, therefore, means that it can be incorporated in this totality. What cannot be incorporated is rejected as incorrect. The alternative to rejection [is to modify the system]. . . . The definition of "correct" and "incorrect" proposed here departs from that customary among the "Vienna Circle," which appeals to "meaning" and "verification." In our presentation we confine ourselves always to the sphere of linguistic thought. (Neurath 1931a/1959, 291; emphasis in original)

11 This issue, the realism/antirealism imbroglio, has been debated since the birth of philosophy, and remains at an "impasse" (Leplin 2000, 393). In barest outline, realism is "commonsensical" (Miller 2002, 13), and although it comes in several varieties (Horwich 2004b), its basic tenets maintain that facts of the world exist, and those facts do not owe their existence to our ability to appreciate them, or even the possibility of knowing them. Following Putnam, philosophers more broadly

refer to this belief as "metaphysical realism." "Scientific realism" draws from that reservoir and refers to a precise position concerning how a scientific theory is to be understood and what scientific activity accomplishes. A simplified definition states that the picture science offers of the world *corresponds* to reality and is therefore true and faithful, so that the entities postulated *really exist as described.* Plainly, *scientific realism* holds that

> successful scientific research is knowledge of largely theory-independent phenomena and that such knowledge is possible (indeed actual) even in those cases in which the relevant phenomena are not, in any non-question-begging sense, observable. . . . [S]ubject to a recognition that scientific methods are fallible and that most scientific knowledge is approximate, we are justified in accepting the most secure findings of scientists "at face value." (Boyd 2002)

In its simplest form, truth for realists corresponds to a human-independent reality. Further, realists assert that science provides an increasingly successful, ever more accurate, asymptotically true, representation of nature. They believe that a universal objectivity is attainable and provides an effective means to discover truths about the world. In short, the realist maintains that "science aims to give us, in its theories, a literally true story of what the world is like; and acceptance of a scientific theory involves the belief that it is true" (van Fraassen 1980, 8). Although science is always changing, and both facts and theories are constantly being modified or discarded, the representation of reality at any given moment is presumably a closer and more reasonable approximation of some ultimate state than previous scientific theories offered. So, while science enjoys no resting place, truth draws ever closer towards defining reality more clearly and completely.

In contrast to the realist, a more skeptical "antirealist" hedges her bets. Despite the assertions of the commonsensical position of the realist, profound philosophical conundrums concerning the relation of mind with the material world arise from the antirealist challenge. While the realist maintains that the world exists independent of human mind, the antirealist asserts that such claims are meaningless because the world can *be known only* cognitively (through mind functions). Our manner of perceiving the world and acting in it depend on the particular character (viz. biology) of the human mind, *and* that the world exists for us (i.e., can be known) as defined by those faculties of knowing. Accordingly, the picture offered by science at any moment in history is a product of the mind *and* nature; more, the mind does not manufacture the world, but rather "the mind and the world jointly make up the mind and the world" (Putnam 1981, xi). Indeed, reality itself is only what human cognition knows, or might know, and this allows the constructivist wedge to break the positivist ideal of objectivity and the corresponding reality so described. Note however, that the argument is not over *reality*, for almost all agree that electrons exist, but rather on *how* we know and whether the mode of knowing determines what *is*.

Antirealism, like realism, is not a single philosophy, but rather captures a variety of critiques opposing the cluster of opinions that comprise realism. A schematic approach would divide the matter as follows: ontologically, antirealism does not dispute the existence of things. Simply, antirealists see an irredeemable conflict between the autonomy of facts (that facts exist independent of us) and their accessibility (the possibility of our gaining knowledge of them) (Horwich 2004a). Epistemologically, the scientific antirealist asserts that the paradigm of *knowledge* is on *observed* facts, which are dependent on human cognition. Such facts may be reformulated and used as theoretical terms, but once an attempt is made to extend such knowledge into a distinct realm of unobservable facts, insuperable obstacles arise, "for how could we ever recognize such facts, or even so much as comprehend them?" (Horwich 2004a, 35). In other words, "theoretical facts" are those that are postulated but not observed.

The degree of epistemological agreement between individual knowers must be very high (because of adaptive evolution, the commonality of language, and the overwhelming evidence of practice), and based on the pragmatic results of the laboratory (i.e., the ability to manipulate nature), it is clear that reality is there to be known, but the issue is *how*. Presumably critical judgment is tempered by human experience bumping into nature and accommodating itself to those realities: "I think it's important for pragmatists to say that the fact there aren't any absolutes of the kind Plato and Kant and orthodox theism have dreamt doesn't mean that every view is as good as every other. It doesn't mean that everything is now arbitrary, or a matter of the will to power, or something like that. That, I think has to be said over and over again" (Rorty 2002, 375). One might then "accept the well established theories of science (even about the unobservable) as (probably) true, but that this should not be understood as accepting the 'metaphysical realist' (Putnam's term) view that the statements which constitute those theories correspond to reality" (Boyd 2002).

Accordingly, truth is derived from the best application of our cognitive functions. The *concepts* we assign to those truth statements comprise the constructivist domain, for the standing of *truth* (final, contingent, deflationist, whatever) constitutes the ongoing practice (or problem) of science (see below). Taking Putnam as representative of this pragmatic stance, the entire enterprise is dependent on an epistemic notion of truth. His argument holds that the realist, in stating the truth conditions for sentences (scientific propositions, theories), says nothing about *knowing* whether those conditions are satisfied because even the best confirmed theories may still be false. Thus he argues that Truth is "some sort of [idealized] rational acceptability" (Putnam 1981, 49) or essentially an epistemic notion.

Even realists agree that scientific theories can only approach full descriptions of reality, never fully capture it, and then they must struggle with the notion of approximate truth (also called verisimilitude). However, realism cannot offer any coherent sense of a theory's approximate truth. For example, a theory may be approximately true, but still inaccurate in those areas where it can be tested. And conversely, over the course of history, we have witnessed theories (e.g., Newtonian optics) that have been successful, but that we now know to be fundamentally flawed. Thus Putnam and like-minded critics sidestep truth in any final sense and argue a much more modest position, namely that scientific theorizing aims to save the phenomena. The criteria are pragmatic: all that can be claimed is what we might observe, and thus a pragmatic position begins by granting that science has indeed achieved success, whether assessed by predictive or manipulative control of nature, or by the precision and parsimony of the descriptions of natural phenomena.

A direct examination of theory testing allows comparative judgments about the reliability of various methods of inquiry and of various theories, and thus an epistemic analysis of the methods of theory testing is required, in other words, an analysis of how the winnowing process operates (Laudan 1984b). The critical restrictions imposed on naive realism leave certain modest philosophical constraints on expectations of science. Moving from a theory's accountability or proximity to truth to the laboratory of experiment, pragmatism again rules (Hacking 1983, 1984). At the level of experimental practice, scientific realism is unavoidable, for there realism is not about theories and truth, since the experimenter need only be a realist about the entities used as tools.

This view draws on various understandings of truth, which fall along a continuum between *robust* theories of truth and *deflationary* ones (Lynch 2001, 4–5). The former assume that truth is an important *property* that requires some explanation. Those who embrace the robust agenda are concerned with the standing of truth as a measure of objectivity and the capacity to capture reality, specifically reality as a realist construes it. In contrast, the deflationist holds that truth has no essential property, and indeed, there is no single robust property or underlying nature to characterize it. So, instead of searching for such a property, called "truth," the deflationist would argue that truth should be regarded as fulfilling an epistemological function (Horwich 2004a; Lynch 1998, 112–13; Armour-Garb and Beall 2005).

Even opponents of realism accept the "certified results of science," and whether realist or antirealist, each must "accept the results of scientific investigations as 'true,' on par with more homely truths. . . . and call this acceptance of scientific truths the 'core position'" (Fine 1986, 128). The antirealist may add certain methodological strictures and caveats to the core position, thus qualifying the truth claims, while the realist will proclaim that the picture is *really* true, but in the end, each rests with some confidence that truth has a best approximation. Again, the difference lies in *how* truth is grounded: for the realist, truth is grounded in some ultimate reality, while the antirealist cautiously claims that the aims of science can be well served without encumbering it with truth criteria that cannot be met. In other words, instead of proclaiming a theory to be true, this "modest realist" would simply display it and enumerate its virtues, which may include the empirical values of adequacy, simplicity, comprehensiveness, coherence, predictability, and so forth (van Fraassen 1980). Thus the scientist is never confronted with a complete theory that need not be susceptible to truth claims of the sort demanded by the scientific realist.

The realism/antirealism debate finds adherents aligned in various permutations of truth theories, but for simplicity's sake, the antirealist will typically reject a correspondence theory of truth. For instance, consider Putnam, who traces his position to Kant, of whom he writes,

> We can view him as rejecting the idea of truth as correspondence (to a mind-independent reality) and as saying that the only sort of truth we can have an idea of, or use for, is *assertibility* (by creatures with our rational natures) *under optimal conditions* (as determined by our sensible natures). Truth becomes a radically epistemic notion" (1983, 210; emphasis in original).

Putnam continued to modify his views (1981; 1983; 1990; 1994), but essentially affirms our common picture of the world and the everyday language by which we deal with it without the conceits of metaphysical realism, and yet holds that criteria of objectivity and truth may still be met, again by consensus. Putnam discards the two traditional avenues that sought to overcome the realism problem. One offers the mind access to "forms," or direct access to the things-in-themselves, thus obviating the problem of correspondence. The other strategy is to postulate a built-in structure of the world, a set of essences, allowing a correspondence between the signs and their objects. Putnam would offer instead

> a species of pragmatism . . . "internal" realism: a realism which recognizes a difference between "p" and "I think that p," between being right, and merely thinking one is right without locating that objectivity in either transcendental correspondence or mere consensus. (Putnam 1983, 225–26)

Or, as he wrote elsewhere:

> The time has come for a moratorium on the kind of ontological speculation that seeks to describe the Furniture of the Universe and to tell us: what is Really There and what is Only a Human Projection, and for a moratorium on the kind of epistemological speculation that seeks to tell us the One Method by which all our beliefs can be appraised. (Putnam 1990, 118)

As an antirealist he is not arguing against the reality of the world, its is-ness, but rather for a realism that refers to our limited ways of knowing that world; limited by the cognitive structures of the mind; knowing understood as an epistemic category. The mind has faculties of cognition, it fashions machines to extend those faculties, and in the process, various methods evolve to find coherence in those methods, with the goal of establishing predictability and mastery over nature. The success of science is precisely in this ability to determine phenomena and/or establish mechanisms for them, and finally to control them. That is enough.

12 See n. 11 above.

13 The problem of natural kinds was well summarized by William James in 1902:

> [Nature] is a vast *plenum* in which our attention draws capricious lines in innumerable directions. We count and name whatever lies upon the special lines we trace, whilst the other things and the untraced lines are neither named nor counted. There are in reality infinitely more things "unadapted" to each other in this world than there are things "adapted;" infinitely more things with irregular relations than with regular relations between them. But we look for the regular kind of thing exclusively, and ingeniously discover and preserve it in our memory. It accumulates with other regular kinds, until the collection of them fills our encyclopedias. Yet all the while between and around them lies an infinite anonymous chaos of objects that no one ever thought of together, of relations that never yet attracted our attention. (James 1902/1987, 394)

14 A shared philosophy underlying characterizations of science and its broader cultural ethos has been notably applied in the twentieth century, for example, (1) logical

positivism and the Bauhaus School of architecture and design during the 1920s and 1930s (Galison 1990), and (2) postmodern indeterminancy to physics and biology (Forman 1971; Griffin 1988). As another example, Paul Forman regards the emergence of quantum mechanics as a particular product of Weimar Germany's *intellectual* climate (1971); a typical physicist would find such a thesis preposterous. In addition, Newtonian physics has been interpreted as emerging from the distinctive political climate of seventeenth-century British political culture (Jacob 1976). A perhaps less radical example considers the social behavior of apes, where gender differences seem operative: Haraway showed that generally, male primatologists see aggressive dominance patterns as prominent, while women scientists observe cooperative relationships as framing social behavior (Haraway 1989a). Such interpretations have, of course, received negative critiques (e.g., Atlan 1993; Tauber 1995).

15 Current constructivist studies go well beyond *The Social Construction of Reality*, written by Peter Berger and Thomas Luckman (1966), wherein they attempted to develop a systematic theory for the sociology of knowledge, and thereby identify its deep penetration into all forms of sociological analyses. "Reality" and "knowledge" fall between philosophy and common sense in the sociological context, for here the concern has shifted from questions of *what* is real or true to a focus on what *counts* as real or true, how validation is attained, and why views of reality and knowledge differ from one culture to another or from one historical moment to the next.

16 See n. 11 above.

17 This orientation on directly assessing scientific practice is not identified with "externalism," for the scientist is acknowledged as operating within a methodological tradition that incorporates theoretical skills that "represent a set of vested interests *within* the scientific community" (Shapin 1982). Reviewed by Gooding, Pinch, and Schaffer (1989); Golinski (1990); Pickering (1992); and Jasanoff et al., (1995), the classic work in this regard, *Genesis and Development of a Scientific Fact*, by Ludwig Fleck (1935/1979), is acknowledged as a seminal inspiration of Kuhn's *Structure* (1962, 1970). Fleck's work is also considered to be the foundational opus for current sociological descriptions concerning the manufacture of a kind of local knowledge strongly dependent upon the practices employed in its making.

18 These empirical orientations partake in pragmatism's distinctive interpretation of *understanding* and *explanation*. As Robert Brandom describes, eighteenth-century enlightenment reason would explain a phenomenon by:

> showing why what *actually* happened *had* to happen that way, why what is actual is (at least conditionally) *necessary*. By contrast for the new pragmatist enlightenment, it is possible to explain what remains, and is acknowledged as, *contingent*. . . . It is not just that we cannot be sure that we have got the principles right. For the correct principles and laws may themselves change. The pragmatists endorse a kind of *ontological fallibilism* or *mutabilism*. . . . The more general philosophical lessons the pragmatists drew from science for an understanding of the nature of reason and its central role in human life accordingly sought to comprehend intellectual understanding as an aspect of effective agency, to situate knowing *that* (some claim is true) in the larger field of knowing *how* (to do something). The sort of explicit reason that can be codified in principles appears as

> just one, often dispensable, expression of the sort of implicit intelligence that can be exhibited in skillful . . . practice—flexible, adaptable habit that has emerged in a particular environment, by selection via a learning process. (Brandom 2004, 2–3; emphasis in original)

Brandom advocates here a reliance on a broad naturalistic epistemology coupled to experience. He is referring to experience not as some "input" to the process of learning, but rather seeing how experience is the process of learning, makes knowledge "an aspect of agency, a kind of doing. Making, not finding, is the genus of human involvement with the world" (5).

Chapter 4

1 To emphasize that the field of science studies fell into two groups, Philip Kitcher (1998, 43) appropriately distinguished one group, radical critics (e.g., Latour and Woolgar 1979; Harding 1986; Woolgar 1988a, 1988b; Pickering 1995), from a more conservative group (e.g., Rudwick 1985; Giere 1988; Hull 1988; Longino 1990; Galison 1992; Dupré 1993; Kitcher 1993). He further maintained that the rabid rebuttal from defenders of science usually failed to distinguish their targets and thereby ignored the substantive contributions made by the latter assembly of scholars and their fellow travelers.

2 An obvious case of extrapolation of the everyday into scientific language and models is the use of metaphor: scientists must work with an implicit vagueness in any groping towards the unknown, and they must borrow from their language and everyday experience to articulate ill-formed models and opaque order of observations. Such metaphoric formulations are expected in theory construction (Ortony 1993, 447–560), and they may be either useful or restrictive. (For examples of the use of metaphor in scientific discourse see Tauber 1994, 2005b; Fox Keller 1995; Cuddington 2001).

3 Sokal recently published a detailed annotation of the parody and his intellectual reflections on the incident and its meaning, a work that includes a vast listing of relevant literature and a spirited defense of what he calls "modest scientific realism" (Sokal 2008).

4 Unique problems arise in writing the history of near-contemporary events and highlight the tensions inherent in a historiography that seeks some distance between the critic and her object of inquiry. When I published a study with Scott Podolsky on the first applications of molecular biology to immunology, We discovered the difficulties of remaining above the fray (Podolsky and Tauber 1997). Our *The Generation of Diversity* is the last of a trilogy on the history and development of immunology's fundamental theory. Unlike its companion volumes (Tauber and Chernyak 1991; Tauber 1994), this last volume profited from interviews with the principal scientists involved in the elucidation of antibody manufacture (an important biological question concerning gene rearrangement and the mechanism of antibody diversification), and we were most pleased to receive their respective insights to help flush out the hidden details of the story. When we sent the manuscript in its penultimate form to the principal investigators for comment, we received some unexpected responses. From those who were given more credit than heretofore

bestowed, we obtained detailed and thoughtful replies. But from the three alpha competitors for the Nobel Prize—Susumu Tonegawa, Leroy Hood, and Philip Leder—we received no constructive responses, despite repeated inquiries concerning specific issues. Hood and Leder declined comment, and Tonegawa, who alone won the coveted award in 1984, told me in the last of several conversations that he would not deign to discuss the manuscript, for it was "worthless," by which he explained that historians could not delve into such matters. When I pressed to know whether there were factual errors or interpretations that might be argued differently, he simply blistered, "You have no idea what I was thinking!" And when I asked him to be specific about where we might have made mistakes, he abruptly terminated the discussion. Tonegawa, needless to say, was disheartened (if not embarrassed) by our description of his role in the discovery of antibody synthesis, and although he was cooperative when he thought we would write in support of his own rendition of that history, he abruptly changed posture when he discovered we were critical of his claims and the history he had written of his own accomplishments. My experience is hardly unique (Söderqvist 1997 and n. 5 below).

5 The extent to which the social scientist should partake in the laboratory investigation itself has, not surprisingly, polarized the science studies community. Some, like Bruno Latour and Steve Woolgar (1979), strongly embrace the "naïve" observer approach, attempting to remain detached observers in order to "avoid the pitfalls of noncritical acceptance of the "scientists' point of view" (Löwy 1997, 92). Others argue a different case, admitting the trials and tribulations of "intimacy," but still maintaining that they must practice their task as informed, indeed expert, observers (e.g., Löwy 1997). Although some espouse an objective distancing, a continuum of practice extends from innocent observer to engaged participant, best exemplified by Garfinkel (1977), who has promoted "ethnomethodology." This program seeks to erase the line of separation between the activity and the interpretation. Like Latour and Woolgar, Garfinkel urges the sociologist to carefully examine practice, tools, and daily discourse, that is, to fully assimilate himself into the laboratory to obtain firsthand, primary data for analysis, as opposed to looking at research papers, institutional structures, statistics of large behavior patterns, and so on. But in embarking on a practice-oriented approach, Garfinkel found himself on a slippery methodological slope (one Woolgar [1988a, 1988b] also slid on, even more radically). Beyond Garfinkel's call to simple empiricism about scientists' behavior, he made a more radical proposal suggesting that ethnomethodology might be "hybridized" with other disciplines such as natural sciences, so that instead of reporting on exotic practices of a physicist or geneticist, the "product" of this hybrid research would in fact be a contribution to the original object of research. In other words, the sociologist in examining laboratory practice would—in this integrated approach—become one of the research participants and thereby contribute to the final assessment.

An ironic conclusion falls out of this program, namely that as a sociological project, it must fail. The ethnomethodologist becomes a scientist, like an anthropologist going native, and although a report is forthcoming, it is by its very nature incomplete and in a sense distorting to the actuality. As Michael Lynch describes the imbroglio:

> The ethnomethodologist would not "go in" to the discipline studied in order to "come back" with a cognitive map or other representation of

> the culture, since no map would be sufficiently complete to recover the scenic details implicated by a competent reading of the map's semiotic features. Short of delivering an entire constellation of details that make up the practical worksite, the only uniquely adequate "news" that could be delivered to the professional sociologists would be an apology to the effect, "You would have had to have been there." (1993, 276)

This is not a unique observation. For instance, Brian Baigrie writes, "We can consult scientific treatises and articles, laboratory notebooks, tables of data, etc., in order to reconstruct some scientific achievement, but *there is no simple historical analogue to practice*" (1995, 106; emphasis in original).

This recognition raises severe problems for doing science studies, and although this is not my specific concern, at the very least, according to Garfinkel (1977), it seems that the historian/sociologist *qua* science studies scholar has forfeited his autonomous professional perspective in favor of going native. Little imagination is required to appreciate how such a proposal was received by natural scientists! The contemporary science historian's or the sociologist's thirst for immersing herself into laboratory practice as a sort of participant observer might be understood as a search for an epistemological "immediacy," in contrast to the theory-driven sociological analysis to which he or she is reacting. This methodological debate has a strong appeal in the social sciences, and represents a long struggle between "idealism" and "observation" that divides these disciplines in so many ways. A second element also contributes to the sociologists' fervor: going beyond Garfinkel, some, like Forman (1997, 198–201), argue that expanded appraisals offer a moral opportunity that should not be squandered. Forman regards an "activist" attitude as a manifestation of a general sense of "responsibility" (192–201). Echoing Feyerabend (1978), he sees science as simply too important to be left to the scientists themselves, and thus critics of science cannot sit in some neutral niche of insular objectivity, but instead they must actively engage the world of technoscience in the moral economy of use. Ilana Löwy (1997, 103–5) similarly picks up this theme in citing Snow's (1959/1964) call for a class of interpreters to mediate the growing gulf between technoscience and the illiterate public, again because of the dire consequences of not assuming responsibility. The problem is, of course, that no consensus on such an adjudicating role has been agreed upon by the scientific community or, for that matter, conferred by the public at large. Finally, Thomas Söderqvist, in his biographical studies, argues that the critic provides some guidance in establishing existential and moral standards for a life in science (2003). His position represents the final turn: historian as priest and moralist, who thereby utterly relieves himself of any objective portrayal of his subject.

6 In addition to seeing science as another species of opinion, Laudan further observes that the extraordinary heterogeneity of the activities and beliefs customarily regarded as scientific suggests the futility of establishing an epistemic version of a demarcation criterion. Since there appear to be no epistemic invariants, one cannot assume their existence, and moreover, even to say there is a boundary problem presupposes the existence of such invariants (Laudan 1996, 221). Laudan summarizes this position: "I am not confident that what we call 'the sciences' have any special set of methodological principles or epistemic credentials that clearly sets them off from

the supposedly 'non scientific' forms of cognition. What I am confident of is the claim that no one, philosopher or sociologist, has yet set out any acceptable account of what cognitive or methodological features demarcate the sciences from the non-sciences" (idem, 189).

7 John McDowell succinctly states this self-critical attitude: "[A]ny thinking . . . is under a standing obligation to reflect about and criticize the standards by which, at any time, it takes itself to be governed. . . . [this] is implicit in the very idea of a shaping of the intellect. . . . This does not mean that such reflection cannot be radical. One can find oneself called on to jettison parts of one's inherited ways of thinking; and [that the] weaknesses that reflection discloses . . . can dictate the formation of new concepts and conceptions. But the essential thing is that one can reflect only from the midst of the way of thinking one is reflecting about" (McDowell 1994, 81).

8 Although success in choosing criteria for scientific verification and prediction may depend on methodological, even technical considerations, there are fundamental concerns that certain phenomena may not be amenable to such scientific analysis. Here I am referring to the dissociation of prediction and theoretical validity, that is, their logical, and in some instances, practical separation. In other words, a prediction may be correct, but its theoretical basis may be false. These are matters, and there are others, that philosophers of science wrestle with, because they seek some logical means by which theories may be judged true, some rigor by which to ascertain the validity of scientific method, and some overarching rationality that explains the predictive success of science. A case in point is evolutionary biology, where the predictive capacities of physics simply do not apply. As a historical discipline, evolutionary theory claims cogency on the parsimony and coherence of its findings, the elucidation of mechanisms that explain them, and the integration of diverse classes of data under a single theory. Thus the criteria for judging success of the life sciences demands emphasis on different values of assessment, or more formally, the relationship of prediction to some encompassing theory.

9 On Richard Rorty's view, the debates about relativism, and the constructivism that supports it, are remnants of epistemologies and metaphysical claims that have outlived their usefulness. He dispensed with any such formal accounts and was content to leave the scientist to pursue nature pragmatically, namely, by the best methods that might be mustered. Thus the truth of any account will simply be the best account available, and that truth is one forged in the crucible of free discourse. In short, truth has no final standing, and indeed cannot (Rorty 1991b).

10 Scientific truth-seeking comprises complex, unformalized procedures that are best characterized as amalgams of many different kinds of methods, rationalities, and their incorporation into a matrix of beliefs that "breathes" in its exchange of ideas and data. So the question arises as to whether such a characterization—antithetical to the positivist tenets that offered advanced industrial societies degrees of certainty unavailable from other knowledge brokers—is beneficial for scientists themselves to understand. Specifically, what salutary benefit might arise if investigators learned the lessons of this self-reflective attitude? Should they be aware that doctrinaire defense of scientific rationality and objectivity rests on contested grounds? Should they be taught that scientific methods are composed of pragmatic adjustments with

certain ideals guiding those accommodations? Should students be advised that "scientific rationality is a myth" (Jardine 2000, 235)?

Chapter 5

1 The boundary issue may be posed from the different perspectives of essentialism and constructivism (Gieryn 1995). Essentialists maintain the possibility and analytic advantage of identifying unique and invariant qualities that set science apart from other occupations, with the larger purpose of explaining its singular achievements; constructivists deny such demarcation principles and instead argue that science, like other intellectual disciplines, is contextually contingent and driven by pragmatic interests of its supporting political culture. Since universal demarcation criteria are not discernable, sociologists of science have settled on examining how, and to what ends, boundaries of science are drawn and defended in practice. From this perspective, boundaries result from a complex interplay of social conventions, both within the "scientific community" (however defined) and from its surrounding culture. In adopting a constructivist attitude, Thomas Gieryn attempts to discredit three essentialist positions: (1) Karl Popper's falsification criteria (1935/1959, 1963), which regards demarcation between science and nonscience as purely an epistemological matter; (2) Kuhn's scientific paradigm concept (1962, 1970) that, despite its attempt to encompass the broadest outline of consensus, remains a nebulous and contested construction of scientific and social elements (see chapter 2); and (3) what is increasingly regarded as the least defensible social theory of science, Robert Merton's normative structure of scientific practice (1973), which most commentators regard as possessing only "surface rules that do not translate into behavior patterns in an immediate and direct way" (Gieryn 1995). This, of course, sounds very much like Wittgenstein's approach to language games, their definition in practice, and the difficulty, if not impossibility, of analytically deciphering the operative rules by which the discourse is governed (Wittgenstein 1953/1968; see chap. 3, n. 8). Beyond this particular application, the entire discipline that falls under the current rubric of sociology of science studies reveals a strong philosophical resonance with a reading of Wittgenstein's philosophy as a "social theory of knowledge" (Bloor 1983), which takes the central topics of epistemology as empirical problems for social science research. Considering the (putative) pivotal role of Wittgenstein, one must ponder the nature of the philosophical bridge that appears to link this epistemology with social constructivism. In this regard, the debate between Michael Lynch and David Bloor (in Pickering 1992) illuminates opposing viewpoints.

2 See chap. 2, n. 3.

3 Although Rorty has been regarded as a bête noir in his iconoclastic attacks on idealized notions of truth and objectivity, I consider his views of their powerful instrumentality compatible with my own. Consider this rather circumspect comment: "The rhetoric of scientific objectivity, pressed too hard and taken too seriously, has led us to people like B. F. Skinner on the one hand and people like Althusser on the other—two equally pointless fantasies, both produced by the attempt to be 'scientific' about our moral and political lives. Reaction against scientism led to attacks on natural science as a sort of false god. But there is nothing wrong with science,

there is only something wrong with the attempt to divinize it . . ." (Rorty 1991a, 33–34).

4 The term "bio centric" appeared approximately a century ago: R. Meldola in *Nature* (January 5, 1899) wrote, "In brief, there has arisen a set of ideas which are even broader than 'anthropocentric', and which might fairly he designated *biocentric*." In 1913, L. J. Henderson charted those limits in *Fitness and Environment*: "The biologist might now rightly regard (the universe in its very essence as biocentric" (312). (Quoted from *Oxford English Dictionary, Suppl. A–G*, 1993, 265). This orientation follows an ethical course in the environmentalist movement, where the entire argument rests on a notion of nature's intrinsic value. Erazim Kohák offers a representative statement of this orientation:

> Every living being, in its strenuous effort to remain alive . . . testifies that its own life is a value for it. . . . The prereflectivity given rule of all life is that life is a value for itself- and as such, a value in itself, internally, quite independently of the existence or of the acts of any other being whatever. *Life is good in itself because It is good for itself. Wherever there is life, there is value.* That is the point: a bio-centric cosmos is not a value-neutral one. (Kohák 1998a, 299; emphasis in original)

This perspective arises from Kohák's phenomenological philosophy, where he maintains that we remain within a meaningfully ordered, value-indexed world, and, more to the point, this is not simply a result of human reflection, but is constituted by life as such. Kohák makes clear in many places in his oeuvre that our ethical relationship with nature is determined by a direct correspondence of our phenomenological encounter, not by any other mediating function such as scientific knowledge. For instance, he wrote regarding "nature as experience":

> Here let us stress sharply: this is not a matter of dealing with nature "as we interpret it" in contrast with nature "as it really is." Nature as experience is how nature really is. . . . That is why, as the starting point of this inquiry, we make neither scientific theory nor the art of argumentation but rather a descriptive phenomenology of the ways humans experience nature . . . and how they experience nature's distress of which we speak as the ecological crisis. (Kohák, 1998b, 258)

I might agree with Kohák's overall position for different reasons, but my discussion is only peripherally concerned with the ethics per se.

5 As Luc Ferry has observed of the relationship between ecological science and values, "We are witnessing the development of the idea that knowledge of the secrets of the universe or of biological organisms endows those who possess it with a new form of wisdom, superior to that of mere mortals. But it is probably in the area of ecology that the feeling that the natural sciences will deliver ready-made teachings applicable to ethics and politics seems to be most confidently asserted" (1995, 84).

6 The status of teleology is perhaps best presented by Jacques Monod in *Chance and Necessity*, where he explicitly id entifies the modern biologist's embarrassment, or what he calls biology's "epistemological contradiction" or biology's "central problem" (1971, 21–22). According to Monod, the contradiction concerns the two ways of knowing that the scientist must employ to study the organic realm. These

Monod calls "objectivity" and "teleonomy." Teleonomy was coined thirty-five years ago to acknowledge end-seeking function without a superimposed design (Tauber 1998). For Monod, teleonomy was employed to designate that organisms achieve their goals *mechanically*, like a computer, with the genetic program offering the necessary direction. Only evolutionary necessity provided the organization for such function, for natural selection was ultimately responsible for writing this program. Bereft of the design connotations of teleology, teleonomy was relegated to simply a vestige of a descriptive biology, eventually (and optimally) to be replaced by a purely mechanical account. Teleonomy, however, does not adequately address the problem of teleology, for while it shoves teleology back to an evolutionary mechanism, the story of adaptation does not account for the descriptive use biologists make of "goal-seeking" function. In short, teleonomy is teleology in disguise.

Whether Monod's open-ended definition of program was warranted or only served as a fanciful metaphor was the subject of dispute. Early critics like Erwin Schrödinger, writing at the time of the birth of cybernetics (1944/1967, 22), noted the difficulty of accounting for a self-written code, which suffers a critical short circuit: control and the controlled have been conflated. Later critics argued that the synthetic merger of information and program worked to introduce again the "argument by design" (Oyama 1985), or that the homunculus had been effectively reinscribed into the gene (Fox Keller 1992). The sleight of hand is accomplished by making a computer analogous to an organism. As Atlan and others have observed, the function and goals of computers are externally prescribed, whereas organisms generate their own behavior, what he calls, the "self-creation of meaning" (1994). But organisms are unable to bracket meaning, which is intrinsic to what we metaphorically refer to as a program (Ben Jacob, Shapira, and Tauber 2005). On this view, programs and information must be distinguished in the organism.

Debate continues concerning what kind of system is capable of generating its own program—its goals, meaning, and tasks. What indeed is a natural machine? Champions of the cybernetic model drew parallels with self-directing machines (like guided missiles or heat-seeking bombs [Jacob 1974, 253]), yet they recognized that the metaphor had limits. It was left to others to add important critical caveats (e.g., Atlan 1983, 28).

Kant explains in the *Critique of Judgment* (1790/1987) that teleology must serve as a *regulative* principle by which organic mechanisms might be examined and understood. Regulative signifies a working method, a mode of understanding, which we use to understand phenomena. It serves as a cognitive category of our own that is employed to organize phenomena, that is, as a faculty of understanding to organize organic studies. In Kant's formulation, teleological explanations orient scientific inquiry that would discern the mechanism's workings. But, and this is the crucial pre-Darwinian position, we might never discern the ultimate basis of design. Kant believed, as he put it, that there never would be a Newton who might explain how a single blade of grass grows or how it appeared in creation. For them, a final intelligence had conferred organization and design to regulate organic function and behavioral ends.

Darwin offered a mechanism that dispensed with divine design. With natural selection, a process of variation and selection explained the evolution of species, their particular anatomies and physiologies. Simply, since Darwin, teleological explanations as construed from Aristotle to Kant have ostensibly been purged from biology. The natural selection theory of evolution, more specifically, its neo-Darwinian formulation, has declared any intentional interpretation as unnecessary for a mechanistic (which would include a stochastic causality) explanation. In short, neo-Darwinism holds that current evolutionary theory, while still incomplete and changing in light of new findings and interpretations, reasonably accounts for the current complexity and phylogenetic history of the organic world. Nevertheless, *telos*—perhaps because it has such a sordid history in biology (Mayr 1982, 1992), and more particularly due to its "semantic instability" (having so many interpretations and inferences)—continues to "taint" biology. ("Teleology is like a mistress to a biologist: he cannot live without her but he's unwilling to be seen with her in public" [attributed to J. B. S. Haldane by Pittendrigh in Mayr 1988, 63].)

7 We might just accept this strange marriage between teleological and mechanistic explanations, either because (1) we cannot completely divorce the two, primarily because the levels of function approached by each are disparate, and the science invoked is as yet inadequately developed (Taylor 1970); or (2) teleological explanations are in fact causal despite their future orientation (Wright 1973, 1976). To capture both teleology's causality and its future orientation, Larry Wright's "consequence-etiology" attempts to integrate goal-seeking behavior as occurring (and thus legitimate for biological descriptions) because it has been causally efficacious in the past (1973, 1976). In other words, goal-directed behavior becomes structured by, or better, has become adaptive as a result of the history of the species or organism as it engages its various goals. Intentionality is thus incorporated into the cause by its past efficacy. Note that on this view, teleology and completely mechanistic accounts of goal-directed behavior may coexist. But as already noted, teleology is being used here as a *description* of adaptive behavior, but like other evolutionary accounts, such descriptions do not adequately expose the nature of biological causality independent of functional purpose (criticisms reviewed by Schaffner [1993, 396–99].) Ken Schaffner, although adopting a more sympathetic view towards reductive analyses than Taylor, acknowledges the current heuristic value of teleological explanations as appropriate to this stage of biology (1993, 379). Atlan assumes a similar position (1994). Yet, teleological thinking has also been productive and need not be thought of exclusively as a vestige of a discarded philosophy of biology. A striking historical case in point is that of Elie Metchnikoff, whose revolutionary view of the organism spawned a bitter dispute with German reductionists who accused him of teleological sin, yet his heuristic framework provided the foundation of the new discipline of immunology. (See Tauber and Chernyak [1991] for an historical account of this most illustrative case.) In both Taylor's and Wright's accounts, what makes behavior teleological is that the process in question can be shown to occur *because* it is required for the achievement of the goal state (Lennox 1992, 332). Each claims a heuristic value. Taylor acknowledges the paradox and adopts a wait and see attitude, while Wright falls into an adaptation mode of explanation.

8 Ecological ethics builds, usually implicitly, on the foundations of evolutionary ethics, which argues how human morality arises from our biological heritage. The argument is well stated by Michael Ruse:

> The new scientific claims are as simple as this. We now know that despite an evolutionary process, centering on a struggle for existence, organisms are not necessarily perpetually at conflict . . . in particular, cooperation can be a good biological strategy. . . . Now let us unpack the science. We begin with the general claims about cooperation, or as today's evolutionists . . . like to call it, altruism. . . . Both the theory and the empirical evidence that biological "altruism" is widespread and promoted by natural selection is very secure and well documented. The simple fact of the matter is that . . . one is frequently better off if one decides to accept a cake shared rather than gambling on the possibility of a whole cake but one which might be lost entirely. (1993, 502)

Thus morality on this view is founded on an evolutionary-derived experience common to myriad species and hardly unique to humans. (See Nitecki and Nitecki [1993] for various presentations.) So altruism, like *telos*, becomes a shared biological characteristic with other animals, and just as we are moral creatures because of those biological endowments, so are our animal brethren, and in a short step, a linkage is established between evolution and environmental ethics.

9 An obvious *telos* of self-definition guides this romantic enterprise that is the contemporary environmental movement. One need not read nature to glorify God, to pay homage to creation, or to seek atonement. The religious intent is directed to only one of the discarded Christian mandates, namely, the saving of the human soul, not for a heavenly paradise, but for a heaven here on earth. Accordingly, in the process of exploring our relationship to nature, we define ourselves; indeed, we are renewed in that encounter. In seeking meaning, Christian revelation thus is translated into a personal naturalized religion, one based on the assertion of our own subjectivity, which allows a means of signifying nature and spiritualizing ourselves (Kirschner 1996). Perhaps few today are aware of their indebtedness to Thoreau's Transcendentalism, but that is hardly the measure of its success as an answer for those seeking meaning in a world increasingly alienated from humane values. And more to the point, American Transcendentalism's abiding religious message holds, namely, that salvation rests firmly within our own grasp (or in secular terms, fate is still of our own making). This understanding is clearly expressed in the environmentalist movement and readily extrapolated to the sciences of the environment and beyond.

Conclusion

1 Herbert Marcuse offers a succinct description of Husserl's criticism regarding the separation of science from its philosophical foundation (which in a different guise sounds similar to Whitehead's own concerns originating from a different philosophical orientation altogether):

> The new science does not elucidate the conditions and the limits of its evidence, validity, and method; it does not elucidate its inherent historical denominator. It remains unaware of its own foundation, and it is therefore unable to recognize its servitude. . . . What happens in the developing

> relation between science and the empirical reality is the abrogation of the transcendence of Reason. Reason loses its philosophical power and its scientific right to define and project ideas and modes of Being beyond and against those established by the prevailing reality. I say "beyond" the empirical reality, not in any metaphysical but in a historical sense, namely, in the sense of projecting essentially different, historical alternatives. (Marcuse 1985, 23)

2 Jürgen Habermas picked up Husserl's challenge and deflected it in a criticism that resonates with a pragmatic sensibility. Simply looking at *theory*, Habermas offers a provocative supplement to our earlier consideration of value-free science, which "reminds us that the postulates associated with it [science] no longer correspond to the classical meaning of theory" (Habermas 1971, 303). He begins his own critique with redefining philosophy from its ancient origins:

> The only knowledge that can truly orient action is knowledge that frees itself from mere human interests and is based on Ideas—in other words, knowledge that has taken a theoretical attitude.
>
> The word "theory" has religious origins. The *theoros* was the representative sent by Greek cities to public celebrations. Through *theoria*, that is through looking on, he abandoned himself to the sacred events. In philosophical language, *theoria* was transferred to contemplation of the cosmos. . . . When the philosopher views the immortal order, he cannot help bringing himself into accord with the proportions of the cosmos and reproducing them internally. . . . Through the soul's likening itself to the ordered motion of the cosmos, theory enters the conduct of life. In *ethos* theory molds life to its form and is reflected in the conduct of those who subject themselves to its disciplines. This concept of theory and life in theory has defined philosophy since its beginnings. (idem, 301–2)

Note that science began with this observational stance, but its philosophical antecedents originally coordinated knowledge within a theoretical frame that examined nature for transposition to man's moral universe. Grounded in *theoria*, order thus included all domains of human knowledge, and science offered its own product for that consideration:

> To dissociate values from facts means counterposing an abstract Ought to pure Being. Values are the nominalistic by-products of a centuries-long critique of the emphatic concept of Being to which theory was once exclusively oriented. The very term "values" . . . to which science is supposed to preserve neutrality, renounces the connection between the two that theory originally intended. Thus, although the sciences share the concept of theory with the major tradition of philosophy, they destroy its classical claim. They borrow two elements from the philosophical heritage: the methodological meaning of the theoretical attitude and the basic ontological assumption of a structure of the world independent of the knower. On the other hand, however, they have abandoned the connection of *theoria* and *kosmos*, of *mimesis* and *bio theoretikos*. . . . What was once supposed to comprise the practical efficacy of theory has now

> fallen prey to methodological prohibitions. The conception of theory as a process of cultivation of the person has become apocryphal. (Habermas 1971, 303–4)

Indeed, Habermas goes on to dismiss Husserl's phenomenological "solution" and instead offers a scheme by which, employing three different methodologies—empirical-analytic, the historical-hermeneutical, and the critical sciences—one abandons the search for a renewed *theoria* as Husserl advocated, and instead admits the limits of objectivity and embraces the connection between knowledge and human interest in the absence of an underlying ontology.

Habermas' criticism of science arises from lingering commitments of a "positivist self-understanding of the sciences" and promotes "the insight that the truth of statements is linked in the last analysis to the intention of the good and true life" (idem, 317). This picks up the pragmatist credo, for human-centered vision of knowledge squarely places science in the service of human interest, which ultimately points to signifying the world as a world of human consciousness. Such a vision of science obviously collapses the rigid distinction of facts and values and promotes instead a personal and pragmatic understanding of knowledge-seeking (i.e., stripped of Platonic ideals of truth or a quest in the employ of metaphysical realism). Such an expanded interpretation of science, one regarded within a fuller humanistic construct, would then become a tool of a larger project: Research becomes a form of creative imagination; scientific literacy opens the world to human wonder; facts, contested and endlessly reconfigured and interpreted, become a currency of competing contexts; objectivity and scientific ethics become widely taught social values. From such an understanding, the dynamic interplay of facts and values creates a kaleidoscopic reality, a "hermeneutical realism" (Dreyfus 1991) that demands dynamic interpretations. In short, with such an understanding, science offers a more comprehensive view of reason in its many voices, as well as a better synthesis of experience derived from diverse sources and generating composite meanings. This is the work of creative imagination. Owen Barfield concisely articulates this orientation:

> [Science] insists on dealing with "data", but there shall no data be given, save the bare precept. The rest is imagination. Only by imagination therefore can the world be known. And what is needed is, not only that larger and larger telescopes and more and more sensitive calipers should be constructed, but that the human mind should become increasingly aware of its own creative activity. (1928/1952/1973, 28)

3 In referring to the religious wonder that underlies science, Atlan might well have referred to Descartes musings on the deepest springs of inquiry and thereby acknowledged this quiet, but persistent thesis. In *Passions of the Soul*, Descartes, in describing the basic passions (wonder, love, hatred, desire, joy, and sadness) from which all other emotions are derived, he places "wonder" in a special category. Unlike the other passions, which cause change in the "heart and blood," wonder is neutral, "having neither good or evil as its object, but only the knowledge of the thing we wonder at" (1649/1931, 363). As an emotion, wonder is characterized as a "sudden surprise of the soul which causes it to apply itself to consider with attention

the objects which seem rare and extraordinary" (idem, 362). Indeed, he pronounces wonder as "useful, inasmuch as it causes us to learn and retain in our memory things of which we were formally ignorant" (364). Descartes concludes that "those who have no natural inclination towards this passion are usually very ignorant" (idem), and like the ancient Greeks before him, he advised moderation to balance wonder's uses and its proper effects.

4 As modern science developed its distinctive epistemology, a new metaphysics also emerged. Indeed, it is disingenuous to insist that science has no metaphysics: as a branch of philosophy it has first principles, presuppositions, which dwell in the deep reaches of its conceptual structure (Whitehead 1925). Science's "logical" structure, includes such precepts as (1) the world is material and ordered; (2) we might discern this order by detached empirical observation, neutral rational description, and objective analysis; (3) laws will emerge from this inquiry and they will remain inviolable; (4) why nature corresponds to our human mathematical and objective descriptions remains mysterious, but the empirical product of that method has been highly successful and thus approximates a depiction of the real as truth, and so on. To be sure, the technical product of this methodological logic, and the power of its predictability, points to a new mastery of nature shared by all.

5 Pertinent to our discussion in chapter 2, Weber, who agonized over the role of value in science, sought an intermediate position that is relevant to our present concerns regarding the romantic view of science. He placed strict borders around science to prevent it being used for ideological purposes (1919/1946), but he also recognized that science, like any human activity, was imbued with value and thus with meaning. He dealt with this problem by highlighting how science's intellectual achievement offers personal satisfaction of human enthusiasm and provides the thrill of inspiration, imagination, and ideas. Accordingly, a scientist is not solely a calculator or uninterested observer but engages in a vital, creative activity. To situate science in terms of its humane function rather than its epistemological aspirations or technological applications, Weber referred to "the inward calling for science"; that is, he addressed the broader meaning of the enterprise for its practitioners (idem). He suggested that the defined scope of scientific disciplines seemed restrictive to this wider agenda, but he recognized that scientific imagination drew upon the same creative sources of intuition that inspires art. Thus he attempted to place value squarely in the experience, not in the product of scientific inquiry. Weber can hardly be called a romantic, but his insight complements those who regarded positivist science as forfeiting claims to becoming a universal philosophy.

A second line of criticism of the romantic view, one which puts a holistic, enchanted vision of science squarely into the political framework, accuses the enchanters of potentially serving dangerous ideology. These critics build on a literature that traces Nazism to its romantic roots and pointedly charge those who would imbue science with value as flirting with the distortion of science and its surreptitious use for ideological ends. "The enchanted version of science, looking for 'value in a world of facts' (Kohler 1938/1959), opens up the possibility that any ethical system can be validated by holistic [enchanted] reason. . . . 'The whole is greater than the sum of its parts' provides the extra something that can be shaped to fit any moral purpose" (Kendler 1999, 831). Accordingly, Anne Harrington (1996) argues that Nazism drew on scientific holism, while George Mosse threw his explanatory net over an even wider swath of German romanticism, which of course included nature's enchantment (Mosse 1964). However:

> the best argument against the Nazi version of 'scientific holism' is not that it was based on values . . . but that the Nazis were simply and objectively wrong in their value judgments (and in their scientific judgments as well). There is no warrant whatsoever for depicting Jews as an "inferior race." This is objectively wrong, scientifically *and* morally. (Brinkmann 2005)

Brinkmann is arguing for the Putnam fact/value collapse thesis (Putnam 1982, 2002); for an extended historical elaboration of this view of Nazi science, see Proctor 1988.

6 It is perhaps useful to note that Thoreau joined a general epistemological shift in the cultural meaning of vision. In the aftermath of Kant, the transparency of the subject-as-observer was generally appreciated as "clouded" as "the science of vision was increasingly linked to new insights in physiology and psychology rather than . . . questions about the mechanics of light and optical transmission" (Pick 1997, 188). As a result, earlier models of unified vision were replaced by attempts to understand how the fragmentation of sensory and psychological experience might be cohered, and more deeply, how objectivity might be attained. Linked to these questions were scientific and artistic explorations of the complex relationships between perception, consciousness, and memory so that by the 1850s:

> a sharply defined and securely describable visual field was giving way to a more ambiguous inner space, the confidence in "clear seeing" dissolving in the face of a renewed preoccupation with how we see objects and how we have "insight." It was well noted by a number of commentators that we would surely view the world and our own minds differently if we could only escape conventional codes. Theorists explored what has since become an unexceptional truth: that expectation and precedent weigh upon our minds, informing and indeed making possible our grasp of the visual world. (Pick, 190)

Thus Thoreau was in good company with Goethe, Friedrich Schelling, Johannes Müller, Herman Helmholtz, J. M. W. Turner, and John Ruskin in not only being self-conscious of the world's presentation in this light, but also being self-conscious of how the mind, in its interactions with nature, creates a world of its own design and character.

7 When discussing science and aesthetics, the line dividing psychology and philosophy remains contested. The separation is particularly vexing because the intersection of the discursive languages are incomplete, and even at cross purposes. For example, how is natural beauty recognized? Geometric form and other visual metaphors generally fulfill criteria of form that we "perceive" as beautiful, but whether the appreciation of a phenomenon or form as beautiful is learned (i.e. culturally derived), or in fact fulfills some resonant cognitive function remains a vexing question (Rentschler et al. 1988). Such matters often serve as the discussion of the aesthetic dimension in science. More to the point, despite its centrality, to pose the question in these terms is to ignore the more fundamental philosophical problem this issue evokes, namely the metaphysics of Cartesian selfhood.

8 For Dewey and Putnam, morals are "objective" in the sense that consensus and considered judgment deliberate the better choice of dealing with the world or drawing inferences from it. A Platonic ideal of "objective" or "real" or "true" then is replaced with a pragmatic assessment adjudicated by the rules of human flourishing. Indeed, as Alasdair MacIntyre cogently confirms:

> To be objective, then, is to understand oneself as part of a community and one's work as part of a project and part of a history. The authority of this history and this project derives from the goods internal to the practice. Objectivity is a moral concept before it is a methodological concept, and the activities of natural science turn out to be a species of moral activity. (1978, 37; see also 1984, 56ff.)

Elizabeth Anderson summarizes accordingly:

> Dewey's ethics replaces the goal of identifying an ultimate end or supreme principle that can serve as a criterion of ethical evaluation with the goal of identifying a method for improving our value judgments. Dewey argued that ethical inquiry is of a piece with empirical inquiry more generally. It is the use of reflective intelligence to revise one's judgments in light of the consequences of acting on them. Value judgments are tools for enabling the satisfactory redirection of conduct when habit no longer suffices to direct it. As tools, they can be evaluated instrumentally, in terms of their success in guiding conduct. We test our value judgments by putting them into practice and seeing whether the results are satisfactory—whether they solve the problems they were designed to solve, whether we find their consequences acceptable, whether they enable successful responses to novel problems, whether living in accordance with alternative value judgments yields more satisfactory results. We achieve moral progress and maturity to the extent that we adopt habits of reflectively revising our value judgments in response to the widest consequences for everyone of living them out. This pragmatic approach requires that we locate the conditions of warrant for our value judgments in human conduct itself, not in any a priori fixed reference point outside of conduct, such as in God's commands, Platonic Forms, pure reason, or "nature," considered as giving humans a fixed *telos*. To do so requires that we understand different types of value judgments in functional terms, as forms of conduct that play distinctive roles in the life of reflective, social beings. Dewey thereby offers a naturalistic metaethic of value judgments, grounded in developmental and social psychology. (2005)

9 As I have previously observed:

> The epistemological and moral domains are not easily separated because we integrate them as informed opinion on a complex continuum between the search for "what is" and our aspirations for "what ought to be." On the social playing field, these two philosophical goals meet somewhere beyond their theoretical origins and thereafter cannot be divided again. (Tauber 1999c, 485)

10 My colleague Victor Kestenbaum has written eloquently on the philosophical challenge of making the ordinary, extraordinary in his study of Dewey, who serves as a touchstone for my own effort. Quoting Charles Tomlinson's poem, "Ode to Arnold Schoenberg" ("Meshed in meaning / by what is natural / we are discontented / for what is more"), Kestenbaum observes:

> We are not meshed in society, history, or language but rather in the meanings—tangible and intangible—which they institute and sustain. Further, we are meshed in meaning "by what is natural." The ranges of "natural" here exceed the specifications, themselves inexact, of philosophical naturalism, including Dewey's naturalism. It is not difficult to see, however, that our discontent does not arise from the state of being "meshed in meaning" but from being "meshed in"—as in "hemmed in"—by meaning which is restricted to the natural. We are discontented [desirous] "for what is more." (2002, 24)

11 I have used Thoreau as a foil of this notion of moral agency, for his moral self-awareness guided all of his various projects—his literary career as pastoralist, his quests for religious and economic integrity, as well as his diverse natural history, political, and aesthetic interests. Robert Milder, in writing of *Walden*, astutely notes, that there are two stories, not always congruent, which unfold in Thoreau's writing: the *narrated* story of discovery and renewal (which we commonly attend to) and the *enacted* story of the writer's efforts to adapt himself to the world (Milder 1995, 54–55), or I would say, to create a moral cosmos. Indeed, using Milder's trope, the two stories converge in the various ways Thoreau attempts to create a self-mythology, one fashioned from his unique visions:

> I believe that there is an ideal or real nature, infinitely more perfect than the actual as there is an ideal life of man. Else where are the glorious summers which in vision sometimes visit my brain When nature ceases to be supernatural to a man—what will he do then? Of what worth is human life—if its actions are no longer to have this sublime and unexplored scenery. Who will build a cottage and dwell in it with enthusiasm if not in the elysian fields? (Thoreau 1981, 481)

In fact, Thoreau did "build a cottage and dwell in it with enthusiasm," by moving into his Walden cabin on July 4, 1845, where he practiced his moral vocation in a life of deliberate introspective practice coupled to a rigorous empiricism and engagement with nature. And in that exercise, instead of offering formal philosophical discourse, he enunciated his distinctive moral philosophy.

References

Abrams, M. H. 1953. *The Mirror and the Lamp. Romantic Theory and the Critical Tradition*. Oxford: Oxford University Press.

Adams, H., and L. Searle, eds. 1986. *Critical Theory 1965*. Tallahasee: Florida State University Press.

Albanese, C. L. 1991. *Nature Religion in America: From the Algonkian Indians to the New Age*. Chicago: University of Chicago Press.

Amrine, F., F. J. Zucker, and H. Wheeler, eds. 1987. *Goethe and the Sciences: A Reappraisal*. Dordrecht, NL: D. Reidel Publishing.

Anderson, E. 2005. *Dewey's Moral Philosophy*. In *Stanford Encyclopedia of Philosophy*, ed. E. N. Zalta. http://plato.stanford.edu/entries/dewey-moral/.

Apel, K-O. 1988. *Understanding and Explanation: A Transcendental-Pragmatic Perspective*. Trans. G. Warnke. Cambridge, Mass: MIT Press.

Armour-Garb, B. P., and J. C. Beall. 2005. "Deflationism: The basics." In Armour-Garb and Beall 2005, 1–29.

Armour-Garb, B. P., and J. C. Beall, eds. 2005. *Deflationary Truth*. Chicago: Open Court.

Ashfield, A., and P. de Bolla, eds. 1996. *The Sublime. A Reader in British Eighteenth-century Aesthetic Theory*. Cambridge: Cambridge University Press.

Ashford, N. A., and K. A. Gregory. 1986. "Ethical problems in using science in the regulatory process." *Natural Resources and the Environment, American Bar Association*, 2:13–57.

Ashmore, M. 1989. *The Reflexive Thesis: Writing Sociology of Scientific Knowledge*. Chicago: University of Chicago Press.

Atlan. H. 1983. "Information theory." In *Cybernetics, Theory and Applications*, ed. R. Trappl. New York: Hemisphere Publishing, 9–41.

———. 1993. *Enlightenment to Enlightenment, Intercritique of Science and Myth.* Trans. L. J. Schramm. Albany: State University of New York Press.

———. 1994. "Intentionality in nature. Against an all-encompassing evolutionary paradigm: Evolutionary and cognitive processes are not instances of the same process." *Journal of Theory of Social Behavior* 24: 67–87.

Ayer, A. J. 1952. "The elimination of metaphysics." In *Language, Truth, and Logic,* 33–45. New York: Dover.

———. 1959. *Logical Positivism.* Glencoe, Ill: The Free Press.

Bachelard. G. 1934/1984. *The New Scientific Spirit.* Trans. A. Goldhammer. Boston: Beacon Press.

Baigrie, B. S. 1995. "Scientific practice: The view from the tabletop." In *Scientific Practice,* ed. Jed Z. Buchwald, 87–122. Chicago: University of Chicago Press.

Bambach, C. R. 1995. *Heidegger, Dilthey, and the Crisis of Historicism.* Ithaca: Cornell University Press.

Barbour, I. G. 1997. *Religion and Science: Historical and Contemporary Issues.* San Francisco: Harper and Row.

Barfield, O. 1928/1952/1973. *Poetic Diction: A Study in Meaning.* Wesleyan, Conn.: Wesleyan University Press.

Barnes, B. 1985. *About Science.* Oxford: Blackwell.

Barnes, B., D. Bloor, and J. Henry. 1996. *Scientific Knowledge: A Sociological Analysis.* Chicago: University of Chicago Press.

Bar-On, R., and J. D. A. Parker, eds. 2000. *Handbook of Emotional Intelligence.* San Francisco: Jossey-Bass.

Barrett, R., and R. Gibson, eds. 1990. *Perspectives on Quine.* Oxford: Blackwell.

Bateson, G. 1980. *Men are Grass: Metaphor and the World of Mental Process.* West Stockbridge, Mass.: Lindisfarne Press.

Beauvoir, S. de. 1953. *The Second Sex.* Trans. H. M. Parshley. New York: Alfred Knopf. Also in Marxists Internet Archive. http://www.marxists.org/reference/subject/ethics/de-beauvoir/2nd-sex/introduction.htm.

Beiser, F. C. 2002. *German Idealism. The Struggle Against Subjectivism, 1781–1801.* Cambridge, Mass.: Harvard University Press.

———. 2005. *Schiller as Philosopher. A Re-examination.* Oxford: Oxford University Press.

Ben Jacob, E., Y. Shapira, and A. I. Tauber. 2005. "Seeking the foundations of cognition in bacteria: From Schrödinger's negative entropy to latent information." *Physica A.* 359:495–524.

Ben-Menahem, Y. 2006. *Conventionalism: From Poincaré to Quine.* Cambridge: Cambridge University Press.

Berger, P. L., and T. Luckman. 1966. *The Social Construction of Reality: A Treatise in the Sociology of Knowledge.* Garden City, N.Y.: Doubleday.

Bernard, C. 1865/1927. *An Introduction to Experimental Medicine.* Trans. by H. C., Green. New York: Dover Publications.

Biagioli, M. 1999. *The Science Studies Reader.* New York: Routledge.

Blackett, P. M. S. 1963. "Memories of Rutherford." In *Rutherford at Manchester*, ed. J. B. Birks, 102–13. New York: W. A. Benjamin.

Bloor, D. 1983. *Wittgenstein: A Social Theory of Knowledge*. New York: Columbia University Press.

———. 1991. *Knowledge and Social Imagery*. 2nd ed. Chicago: University of Chicago Press.

Bortoft, H. 1996. *The Wholeness of Nature: Goethe's Way toward a Science of Conscious Participation in Nature*. Hudson, N.Y.: Lindisfarne Press.

Bower, G. H. 1981. "Mood and memory." *American Psychologist* 36:129–48.

Boyd, R. N. 1984. "The current status of scientific realism." In Leplin 1984, 41–82.

———. 2002. *Scientific Realism*. In *Stanford Encyclopedia of Philosophy*, ed. E. N. Zalta. http://plato.stanford.edu/archives/sum2002/entries/scientific-realism/.

Boyd, R. N., P. Gasper, and I. D. Trout, eds. 1991. *The Philosophy of Science*. Cambridge, Mass.: MIT Press.

Brandom, R. B. 2004. "The pragmatist Enlightenment (and its problematic semantics)." *European Journal of Philosophy* 12:1–16.

Brandt, R. B. 1979. *A Theory of the Good and the Right*. Oxford: Oxford University Press.

Braver, L. 2007. *A Thing of the World: A History of Continental Anti-realism*. Evanston, Ill.: Northwestern University Press.

Brinkmann, S. 2005. "Psychology's facts and values: A perennial entanglement." *Philosophical Psychology* 18:749–65.

Bronowski, J. 1956. *Science and Human Values*. New York: Harper and Row.

Brooke, J. H. 1991. *Science and Religion: Some Historical Perspectives*. Cambridge: Cambridge University Press.

Brown, J. R. 2001. *Who Rules in Science? An Opinionated Guide to the Wars*. Cambridge, Mass.: Harvard University Press.

Bruce, R. V. 1987. *The Launching of Modern American Science 1846–1876*. Ithaca: Cornell University Press.

Bulger, R. E., E. Heitman, and S. J. Reiser, eds. 1993. *The Ethical Dimensions of the Biological Sciences*. Cambridge: Cambridge University Press.

Bunge, M. 1996. "In praise of intolerance to charlatanism in academia." In Gross, Levitt, and Lewis 1996, 96–115.

Butterfield, H. 1957. *The Origins of Modern Science*. London: G. Bell and Sons.

Callon, M. 1995. "Four models for the dynamics of science." In Jasanoff, Markle, Petersen, and Pinch 1995, 29–63.

Canguilhem, G. 1989. *The Normal and the Pathological*. Trans. C. R. Fawcett. New York: Zone Books.

Carruthers P., S. Stich, and M. Siegal, eds. 2002. *The Cognitive Basis of Science*. Cambridge: Cambridge University Press.

Cartwright, N. 1983. *How the Laws of Physics Lie*. Oxford: Oxford University Press.

———. 1999. *The Dappled World: A Study of the Boundaries of Science*. Cambridge: Cambridge University Press.

Chadwick, O. 1975. *The Secularization of the European Mind in the 19th Century*. Cambridge: Cambridge University Press.

Chubin, D. E. 1990. "Scientific malpractice and the contemporary politics of knowledge." In Cozzens and Gieryn, 1990, 144–63. Bloomington: Indiana University Press.

Cohen, H. F. 1994. *The Scientific Revolution: A Historiographical Inquiry*. Chicago: University of Chicago Press.

Cohen, R. S. 1974. "Ethics and science." In *For Dirk Struik. Scientific Historical and Political Essays in Honor of Dirk J. Struik*, eds. R. S. Cohen, J. J. Stachel, and M. W. Wartofsky, 307–23. Dordrecht, NL: D. Reidel Publishing.

Cohen, R. S., and M. Neurath, eds. 1983. *Philosophical Papers 1913–1946*. Dordrecht, NL: D. Reidel Publishing.

Cole, S. 1992. *Making Science. Between Nature and Society*. Cambridge, Mass.: Harvard University Press.

Collingwood, R. G. 1940. *An Essay on Metaphysics*. Oxford: Clarendon Press.

Collins, H. M. 1992. *Changing Order. Replication and Induction in Scientific Practice*. Chicago: University of Chicago Press.

Collins, H. M., and T. Pinch. 1982. *Frames of Meaning. The Social Construction of Extraordinary Science*. London: Routledge and Kegan Paul.

Comte, A. 1825/1974. "Philosophical considerations on the sciences and savants." In *The Crisis of Industrial Civilization: The Early Essays of Auguste Comte*, ed. and trans. R. Fletcher, 182–213. New York: Crane, Rusak.

Conant, J. B. 1953. *Modern Science and Modern Man*. Garden City, N.Y.: Doubleday.

Cosslett, T. 1982. *The Scientific Movement and Victorian Literature*. New York: St. Martin's Press.

———. 1984. *Science and Religion in the Nineteenth Century*. Cambridge: Cambridge University Press.

Cozzens, S. E., and T. F. Gieryn, eds. 1990. *Theories of Science in Society*. Bloomington: Indiana University Press.

Croce, B. 1902/1972. *Aesthetic as Science of Expression and General Linguistic*. Trans. D. Ainslie. New York: Farrar, Straus, and Giroux.

Cuddington, K. 2001. "The 'balance of nature' metaphor and equilibrium in population ecology." *Biology and Philosophy* 16:463–79.

Cushing, J. T., C. F. Delaney, and G. M. Gutting, eds. 1984. *Science and Reality: Recent Work in the Philosophy of Science*. Notre Dame: University of Notre Dame Press.

Damasio, A. 1994. *Descartes' Error: Emotion, Reason, and the Human Brain*. New York: G. P. Putnam's Sons.

Darwin, C. 1987. *Darwin's Notebooks. 1836–1844*. Eds. P. H. Barrett, et al. Ithaca: Cornell University Press.

Daston, L. 1994. "Baconian facts, academic civility, and the prehistory of objectivity." In Megill 1994, 37–64.

———. 2000. "Scientific objectivity with and without words." In *Little Tools of Knowledge: Historical Essays on Academic and Bureaucratic Practice*, eds. P. Becker and W. Clark. Ann Arbor: University of Michigan Press, 259–84.

Daston, L. and P. Galison. 2007. *Objectivity*. New York: Zone Books.

Davidson, D. 1980. "Actions, reasons, and causes." In *Essays on Actions and Events*. 3–19. Oxford: Clarendon Press.

Davies, P. C. W., and J. R. Brown. 1986. *The Ghost in the Atom: A Discussion of the Mysteries of Quantum Physics*. Cambridge: Cambridge University Press.

Davidson, K. 2006. *The Death of Truth: Thomas S. Kuhn and the Evolution of Ideas*. New York: Oxford University Press.

Deacon, T. 1997. *The Symbolic Species: The Co-evolution of Language and the Brain*. New York: W. W. Norton.

Dear, P. 2001. "Science studies as epistemography." In Labinger and Collins 2001, 128–41.

Dembski, W. A., and M. Ruse, eds. 2004. *Debating Design: From Darwin to DNA*. Cambridge: Cambridge University Press.

Dennett, D. 1981. *Brainstorms: Philosophical Essays on Mind and Psychology*. Cambridge, Mass.: MIT Press.

Descartes, R. 1649/1931. *The Passions of the Soul*. In *Philosophical Works of Descartes*, vol. 1. Trans. E. S. Haldane and G. R. T. Ross, 331–427. Cambridge: Cambridge University Press.

de Sousa, R. 1987. *The Rationality of Emotion*. Cambridge, Mass.: MIT Press.

de Waal F. 1996. *Good Natured: The Origins of Right and Wrong in Humans and Other Animals*. Cambridge, Mass.: Harvard University Press.

Dewey, J. 1922. *Human Nature and Conduct: An Introduction to Social Psychology*. New York: Henry Holt. Republished Amherst, N.Y.: Prometheus Books, 2002.

———. 1929/1984. *The Quest for Certainty*. Carbondale: Southern Illinois University Press.

———. 1931. "Philosophy and civilization." In *Philosophy and Civilization*. New York: Capricorn Books.

———. 1920/1948. *Reconstruction in Philosophy*. Boston: Beacon Press.

Doel, R. E. 1997. "Scientists as policy makers, advisors, and intelligence agents: Linking contemporary diplomatic history with the history of contemporary science." In Söderqvist 1997, 215–44.

Dreyfus, H. L. 1991. "Heidegger's hermeneutic realism." In *The Interpretive Turn. Philosophy, Science, Culture*, eds. D. S. Hiley, J. F. Bohman, and R. Shusterman, 25–41. Ithaca: Cornell University Press.

Duhem, P. 1906/1954. *The Aim and Structure of Physical Theory*. Trans. P. P. Wiener. Princeton: Princeton University Press.

Dunlap, T. R. 2005. *Faith in Nature: Environmentalism As Religious Quest*. Seattle: University of Washington Press.

Dupré, J. 1993. *The Disorder of Things: Metaphysical Foundations of the Disunity of Science*. Cambridge, Mass.: Harvard University Press.

Editors of *Lingua Franca*. 2000. *The Sokal Hoax: The Sham that Shook the Academy.* Lincoln: University of Nebraska Press.

Egginton, W., and M. Sandbothe, eds. 2004. *The Pragmatic Turn in Philosophy: Contemporary Engagements Between Analytic and Continental Thought.* Albany: State University of New York Press.

Emerson, R. W. 1837/1964. "The present age." In *The Early Lectures of Ralph Waldo Emerson*, vol. 2, 1836–1838, eds. S. E. Whicher, R. E. Spiller, and W. E. Williams. Cambridge, Mass.: Harvard University Press.

———. 1836/1971. "Nature." In *The Collected Works of Ralph Waldo Emerson.* Vol. 1 of *Nature, Addresses, and Lectures*, 1–45. Cambridge, Mass.: Harvard University Press.

Evernden, N. 1992. *The Social Creation of Nature*. Baltimore: The Johns Hopkins University Press.

Farrell, R. P. 2003. F*eyerabend and Scientific Values: Tightrope-Walking Rationality.* Dordrecht, NL: Kluwer Academic Publishers.

Fauconnier, G., and M. Turner. 2002. *The Way We Think: Conceptual Blending and the Mind's Hidden Complexity*. New York: Basic Books.

Faust, D. 1984. *The Limits of Scientific Reasoning*. Minneapolis: University of Minnesota Press.

Fay, B., P. Pomper, and R. T. Vann, eds. 1998. *History and Theory: Contemporary Readings*. Malden, Mass.: Blackwell.

Feldman, N. 2005. *Divided by God: America's Church-State Problem—and What We Should do About It.* New York: Farrar, Straus, and Giroux.

Ferngren, G. B., ed. 2002. *Science and Religion: A Historical Introduction.* Baltimore: The Johns Hopkins University Press.

Ferry, L. 1995. *The New Ecological Order.* Chicago: University of Chicago Press.

Feyerabend, P. 1975. *Against Method.* London: Verso.

———. 1978. *Science in a Free Society*. London: NLB.

Fine, A. 1986. "The natural ontological attitude." In *The Shaky Game: Einstein, Realism and the Quantum Theory*, 112–35. Chicago: University of Chicago Press.

———. 1991. "Is scientific realism compatible with quantum physics?" In *The Philosophy of Science*, eds. R. Boyd, P. Gasper, and J. D. Trout, 529–44. Cambridge, Mass.: MIT Press.

Fink, K. J. 1991. *Goethe's History of Science*. Cambridge: Cambridge University Press.

Fisch, M. 2006. *Rational Rabbis: Its Project and Argument. Journal of Textual Reasoning.* http://etext.virginia.edu/journals/tr/volume4/number2/TR04_02_e01.html.

Fleck, L. 1935/1979. *Genesis and Development of a Scientific Fact.* Chicago: University of Chicago Press.

Forman, P. 1971. "Weimar culture, causality, and quantum theory, 1918–1927: Adaptation by German physicists and mathematicians to a hostile intellectual environment." *Historical Studies in the Physical Sciences* 3:1–115.

———. 1997. "Recent science: Late-modern and post-modern." In Söderqvist 1997, 179–213.

———. 2007. "The primacy of science in modernity, of technology in postmodernity, and of ideology in the history of technology." *History and Technology* 23:1–152.

Foucault, M. 1980. *Power/Knowledge: Selected Interviews and Other Writings, 1972–1977*. New York: Pantheon.

———. 1994. *The Order of Things: An Archaeology of Human Sciences*. New York: Vintage.

Fox Keller, E. 1992. "A vision of the grail." In *The Code of Codes. Scientific and Social Issues in the Human Genome Project*, eds. D. J. Kevles and L. Hood, 83–97. Cambridge, Mass.: Harvard University Press.

———. 1994. "The paradox of scientific subjectivity." In Megill 1994, 313–31.

———. 1995. *Refiguring Life: Metaphors of Twentieth-century Biology*. New York: Columbia University Press.

Frank, P. 1949. *Modern Science and its Philosophy*. Cambridge, Mass.: Harvard University Press.

Friedman, M. 1999. *Reconsidering Logical Positivism*. Cambridge: Cambridge University Press.

———. 2000. *A Parting of the Ways. Carnap, Cassirer, and Heidegger*. Chicago: Open Court.

Fujimura, I. 1992. "Crafting science: Standardized packages, boundary objects, and 'translation.'" In Pickering 1992, 168–211.

Fuller, S. 2001. *Thomas Kuhn: A Philosophical History of Our Times*. Chicago: University of Chicago Press.

Gadamer, H-G. 1981. *Reason in the Age of Science*. Trans. F. G. Lawrence. Cambridge, Mass.: MIT Press.

———. 1990. *Truth and Method*. 2nd rev. ed. New York: Crossroad.

Galaty, D. H. 1974. "The philosophical basis for mid-nineteenth-century German reductionism." *Journal of the History of Medicine and Allied Sciences* 29:295–316.

Galison, P. 1990. "Aufbau/Bauhaus: Logical positivism and architectural modernism." *Critical Inquiry* 16:709–52.

———. 1992. *How Experiments End*. Chicago: University of Chicago Press.

Galison, P., and D. J. Stump, eds. 1996. *The Disunity of Science. Boundaries, Contexts, and Power*. Stanford: Stanford University Press.

Gardner, H. 1993. *Frames of Mind: The Theory of Multiple Intelligences*. New York: Basic Books.

Garfinkel, H. 1977. "When is phenomenology sociological? A panel discussion with J. O'Neill, G. Psathas, E. Rose, E. Tiryakian, H. Wagner, and D. L. Wieder." *Annals of Phenomenological Sociology*, 1–40.

Gergen, K. J. 1994. "The mechanical self and the rhetoric of objectivity." In Megill 1994, 265–88.

Giere, R. N. 1988. *Explaining Science. A Cognitive Approach.* Chicago: University of Chicago Press.

———. 1997. "Kuhn's legacy for North American philosophy of science." *Social Studies of Science* 27:496–98.

———. 1999. *Science Without Laws.* Chicago: University of Chicago Press.

Giere, R. N., and A. W. Richardson, eds. 1996. *Origins of Logical Empiricism, Minnesota Studies in Philosophy of Science* 16. Minneapolis: University of Minnesota Press.

Gieryn, T. F. 1995. "Boundaries of science." In Jasanoff, Markle, Petersen, and Pinch 1995, 393–443.

Gieryn, T. F., and A. E. Figert. 1990. "Ingredients for the theory of science in society: O-rings, ice water, C-clamp, Richard Feynman and the press." In Cozzens and Gieryn, 67–97.

Gilbert, S. F. 1997. "Bodies of knowledge: Multiculturalism and science." In *Changing Life: Genomes, Ecologies, Bodies, Commodities*, eds. P. J. Taylor, S. E. Halfon, and P. N. Edwards, 36–55. Minneapolis: University of Minnesota Press.

Gilbert, W. 1992. "A vision of the grail." In *The Code of Codes. Scientific and Social Issues in the Human Genome Project*, eds. D. J. Kevles and L. Hood, 83–97. Cambridge, Mass.: Harvard University Press.

Goethe J. W. 1790/1989. "The metamorphosis of plants." In Mueller 1989, 31–78.

———. 1794/1988. "The extent to which the idea 'Beauty is Perfection in Combination with Freedom' may be applied to living organisms." In D. Miller 1988, 22–23.

———. 1810/1988. *Theory of Colours.* In D. Miller 1988, 157–298.

———. 1817a/1988. "The influence of modern philosophy." In D. Miller 1988, 28–30.

———. 1817b/1989. "History of the printed brochure." In Mueller 1989, 170–76.

———. 1823/1988. "Significant help given by an ingenious turn of phrase." In D. Miller 1988, 39–41.

———. 1829/1998. *Maxims and Reflections.* New York: Penguin Books.

Golinski, J. 1990. "The theory of practice and the practice of theory: Sociological approaches in the history of science." *ISIS* 81:492–505.

———. 2005. *Making Natural Knowledge: Constructivism and the History of Science.* Chicago: University of Chicago Press.

Gooding, D., T. Pinch, and S. Schaffer, eds. 1989. *The Uses of Experiment. Studies in the Natural Sciences.* Cambridge: Cambridge University Press.

Goodman, N. 1972. "The way the world is." In *Problems and Projects*, 24–32. Indianapolis: Bobbs-Merrill.

Gower, B. 1997. *Scientific Method: An Historical and Philosophical Introduction.* London: Routledge.

Greenberg, D. S. 1967. *The Politics of Pure Science.* New York: New American Library.

———. 2003. *Science, Money, and Politics: Political Triumph and Ethical Erosion.* Chicago: University of Chicago Press.

———. 2007. *Science for Sale: The Perils, Rewards, and Delusions of Campus Capitalism.* Chicago: University of Chicago Press.

Griffin, D. R., ed. 1988. *The Reenchantment of Science: Postmodern Proposals.* Albany: State University of New York Press.

Gross, P. R., and N. Levitt. 1994. *Higher Superstition: The Academic Left and Its Quarrels with Science.* Baltimore: The Johns Hopkins University Press.

Gross P. R., N. Levitt, and M. W. Lewis, eds. 1996. *The Flight From Science and Reason*, vol. 775. New York: New York Academy of Sciences.

Gutting, G. 1983. "Scientific realism vs. constructive empiricism: A dialogue." *Monist* 65:336–49.

———, ed. 2005. *Continental Philosophy of Science.* Malden, Mass.: Blackwell.

Haack, S. 1996. "Concern for truth: What it means, why it matters." In Gross, Levitt, and Lewis 1996, 57–63.

———. 1998. *Manifesto of a Passionate Moderate.* Chicago: University of Chicago Press.

Habermas, J. 1971. *Knowledge and Human Interests.* Trans. J. J. Shapiro. Boston: Beacon Press.

Hacking, I. 1983. *Representing and Intervening. Introductory Topics in the Philosophy of Natural Science.* Cambridge: Cambridge University Press.

———. 1984. "Experimentation and scientific realism." In Leplin 1984, 154–72.

———. 1999. *The Social Construction of What?* Cambridge, Mass.: Harvard University Press.

Hansen, O. 1990. *Aesthetic Individualism and Practical Intellect: American Allegory in Emerson, Thoreau, Adams, and James.* Princeton: Princeton University Press.

Hanson, N. R. 1958. *Patterns of Discovery: An Inquiry into the Conceptual Foundations of Science.* Cambridge: Cambridge University Press.

Haraway, D. 1989a. *Primate Visions: Gender, Race, and Nature in the World.* London: Routledge and Kegan Paul.

———. 1989b. "The biopolitics of postmodern bodies: Determinations of self in immune systems discourse. differences." *A Journal of Feminist Cultural Studies,* 1: 3–43.

Harding, S. 1986. *The Science Question in Feminism.* New York: Cornell University Press.

Harrington, A. 1996. *Reenchanted Science: Holism in German Culture from Wilhelm II to Hitler.* Princeton: Princeton University Press.

Harrison, P. 1998. *The Bible, Protestantism, and the Rise of Natural Science.* Cambridge: Cambridge University Press.

Harvey, C. W. 1989. *Husserl's Phenomenology and the Foundations of Natural Science.* Athens: Ohio University Press.

Heal, J. 1987–1988. "The disinterested search for truth." *Proceedings of the Aristotelian Society* 88:97–108.

Hegel, G. W. F. 1821a/1896. *Philosophy of Right.* Trans. S. W. Dyde. Marxist Internet Archive. http://www.marxists.org/reference/archive/hegel/.

———. 1821b/1952. *Hegel's Philosophy of Right.* Trans. T. M. Knox. Oxford: Oxford University Press.

———. 1821c/1991. *Elements of the Philosophy of Right.* Trans. H. B. Nisbet. Ed. A. W. Wood. Cambridge: Cambridge University Press.

Heidegger, M. 1954a/1977. "The age of the world picture." In *The Question Concerning Technology and Other Essays*, trans. W. Lovitt, 115–54. New York: Harper Torchbooks.

———. 1954b/1977. "Science and reflection." In *The Question Concerning Technology and Other Essays*, trans. W. Lovitt, 155–82. New York: Harper Torchbooks.

———. 1961/1979. *Nietzsche.* Vol. 1, *The Will to Power as Art.* Trans. D. F. Krell. New York: HarperCollins.

Heisenberg, W. 1958a. "The Copenhagen interpretation of quantum theory." In *Physics and Philosophy: The Revolution in Modern Science.* New York: Harper & Row, 44–58.

———. 1958b. "The development of philosophical ideas since Descartes in comparison with the new situation in quantum theory." In *Physics and Philosophy: The Revolution in Modern Science.* New York: Harper and Row, 76–92.

———. 1979. "The teachings of Goethe and Newton on colour in the light of modern physics." In *Philosophical Problems of Quantum Physics*, 60–76. Woodbridge: Oxbow Press.

Hempel, C. G. 1966. *Philosophy of Natural Science.* Englewood Cliffs, N.J.: Prentice Hall.

Hodder, A. D. 2001. *Thoreau's Ecstatic Witness.* New Haven: Yale University Press.

Hollis M., and S. Lukes. 1982. *Rationality and Relativism.* Cambridge, Mass.: MIT Press.

Holmes, F. L. 1974. *Claude Bernard and Animal Chemistry.* Cambridge, Mass.: Harvard University Press.

Holton, G. 1993. *Science and Anti-Science.* Cambridge, Mass.: Harvard University Press.

Hookway, C. 1988. *Quine.* Stanford: Stanford University Press.

———. 2002. "Emotion and epistemic evaluations." In Carruthers, Stich, and Siegal 2002, 251–62.

Hopp, W. 2008. "Husserl, phenomenology, and foundationism." *Inquiry* 51:194–216.

Horkheimer, M. 1947/2004. *Eclipse of Reason.* London: Continuum.

Horwich, P., ed. 1993. *World Changes: Thomas Kuhn and the Nature of Science.* Cambridge, Mass.: MIT Press.

———. 2004a. "Realism and truth." In *From a Deflationary Point of View*, 32–44. Oxford: Clarendon Press.
———. 2004b. "Three forms of realism." In *From a Deflationary Point of View*, 7–31. Oxford: Clarendon Press.
Hoyningen-Huene, P. 1993. *Reconstructing Scientific Revolutions. Thomas S. Kuhn's Philosophy of Science*. Trans. A. T. Levine. Chicago: University of Chicago Press.
Hull, D. 1988. *Science As a Process: An Evolutionary Account of the Social and Conceptual Development of Science*. Chicago: University of Chicago Press.
Hume, D. 1739/1978. *A Treatise of Human Nature*. Oxford: Clarendon Press.
Husserl, E. 1935/1970. *The Crisis of European Science and Transcendental Phenomenology: An Introduction to Phenomenological Philosophy*. Trans. D. Carr. Evanston: Northwestern University Press.
Hylton, P. 1990. *Russell, Idealism and the Emergence of Analytic Philosophy*. Oxford: Clarendon Press.
Irzik, G., and T. Grünberg. 1995. "Carnap and Kuhn: Arch enemies or close allies?" *British Journal of Philosophy of Science* 46:285–307.
Jacob, F. 1974. *The Logic of Life*. New York: Pantheon Press.
Jacob, M. C. 1976. *The Newtonians and the English Revolution 1689–1720*. Ithaca: Cornell University Press.
James, W. 1902/1987. *The Varieties of Religious Experience*. New York: The Library of America.
Jammer, M. 1974. *The Philosophy of Quantum Mechanics: The Interpretations of Quantum Mechanics in Historical Perspective*. New York: John Wiley & Sons.
Jardine, N. 2000. *The Scenes of Inquiry: On the Reality of Questions in the Sciences*. Oxford: Clarendon Press.
Jasanoff, S. 1998. *The Fifth Branch: Science Advisors as Policymakers*. Cambridge, Mass.: Harvard University Press.
Jasanoff, S., G. E. Markle, J. C. Petersen, and T. Pinch, eds. 1995. *Handbook of Science and Technology Studies*. Thousand Oaks, Calif.: Sage Publications.
Jonas, H. 1963. *The Gnostic Religion. The Message of the Alien God and the Beginnings of Christianity*. Boston: Beacon Press.
———. 1984. *The Imperative of Responsibility*. Chicago: University of Chicago Press.
Judson, H. F. 2004. *The Great Betrayal: Fraud in Science*. New York: Harcourt.
Kant, I. 1787/1998. *Critique of Pure Reason*. Trans. P. Guyer and A. W. Wood. Cambridge: Cambridge University Press.
———. 1790/1987. *Critique of Judgment*. Trans. W. S. Pluhar. Indianapolis: Hackett.
Kendler, H. 1999. "The role of value in the world of psychology." *American Psychologist* 54:828–35.
Kestenbaum, V. 2002. *The Grace and Severity of the Ideal: John Dewey and the Transcendent*. Chicago: University of Chicago Press.
Kincaid, H. 1990. "Defending laws in social sciences." *Philosophy of Social Science* 20:56–83.

Kincaid, H., J. Dupré, and A. Wylie. 2007. *Value-free Science? Ideals and Illusions.* Oxford: Oxford University Press.

Kirschner, S. R. 1996. *The Religious and Romantic Origins of Psychoanalysis.* Cambridge: Cambridge University Press.

Kitcher, P. 1993. *The Advancement of Science. Science without Legend, Objectivity without Illusions.* New York: Oxford University Press.

———. 1998. "A plea for science studies." In Koertge 1998, 32–56.

Knight, D. 1986. *The Age of Science: The Scientific World-view in the Nineteenth Century.* Oxford: Basil Blackwell.

Knorr Cetina, K. 1999. *Epistemic Cultures: How Science Makes Knowledge.* Cambridge, Mass.: Harvard University Press.

Koertge, N., ed. 1998. *A House Built on Sand. Exposing Myths about Science.* New York: Oxford University Press.

Kohák, E. 1998a. "The ecological dilemma: Ethical categories in a biocentric world." In *Philosophies of Nature: The Human Dimension*, eds. R. S. Cohen and A. I. Tauber, 293–301. Dordrecht, NL: Kluwer Academic Publishers.

———. 1998b. "Varieties of ecological experience." In *Philosophies of Nature: The Human Dimension,* eds. R. S. Cohen and A. I. Tauber, 257–71. Dordrecht, NL: Kluwer Academic Publishers.

Kohler, W. 1938/1959. *The Place of Value in a World of Facts.* New York: Liverwright.

Kolakowski, L. 1968. *The Alienation of Reason: A History of Positivist Thought.* Trans. N. Guterman. Garden City, N.Y.: Doubleday.

Kornberg, A. 1995. *The Golden Helix: Inside Biotech Ventures.* Sausalito, Calif.: University Science Books.

Koyré, A. 1957. *From the Closed World to the Infinite Universe.* Baltimore: The Johns Hopkins University Press.

Kragh, H. 1987. *An Introduction to the Historiography of Science.* Cambridge: Cambridge University Press.

Kremer, R. L. 1990. *The Thermodynamics of Life and Experimental Physiology 1770–1880.* New York: Garland Publishing.

Krimsky, S. 2003. *Science in the Private Interest: Has the Lure of Profits Corrupted Biomedical Research?* Lanham, Md.: Rowan and Littlefield.

Kuhn, T. S. 1962. *The Structure of Scientific Revolutions.* Chicago: University of Chicago Press.

———. 1970. *The Structure of Scientific Revolutions.* 2nd ed. Chicago: University of Chicago Press.

———. 1991. "The road since *Structure.*" *PSA* 2:3–13.

———. 1992. "The trouble with the historical philosophy of science." Cambridge, Mass.: Department of the History of Science, Harvard University.

Kukla, A. 2000. *Social Constructivism and the Philosophy of Science.* London: Routledge.

Kusch, M. 2002. *Knowledge by Agreement: The Programme of Communitarian Epistemology.* Oxford: Clarendon Press.

Labinger, J. A., and H. Collins, eds. 2001. *The One Culture? A Conversation about Science*. Chicago: University of Chicago Press.

Lacey, H. 1999. *Is Science Value Free? Values and Scientific Understanding*. London and New York: Routledge.

Lafollette, M. C. 1996. *Stealing into Print: Fraud, Plagiarism, and Misconduct in Scientific Publishing*. Los Angeles: University of California Press.

Lakatos, I. 1970. "Falsification and the methodology of scientific research programmes." In *Criticism and the Growth of Knowledge,* eds. I. Lakatos and A. Musgrave, 91–196. Cambridge: Cambridge University Press.

Latour, B. 1987. *Science in Action. How to Follow Scientists and Engineers through Society*. Cambridge, Mass.: Harvard University Press.

———. 1988. *The Pasteurization of France*. Cambridge, Mass.: Harvard University Press.

———. 1993. *We Have Never Been Modern*. Cambridge, Mass.: Harvard University Press.

———. 1999. *Pandora's Box: Essays on Reality of Science Studies*. Cambridge, Mass.: Harvard University Press.

Latour, B., and S. Woolgar. 1979. *Laboratory Life: The Construction of Scientific Facts*. Princeton: Princeton University Press.

Laudan, L. 1984a. *Science and Values: The Aims of Science and Their Role in Scientific Debate*. Berkeley: University of California Press.

———. 1984b. "Explaining the success of science: Beyond epistemic realism and relativism." In Cushing, Delaney, and Gutting 1984, 83–105.

———. 1990a. *Science and Relativism: Some Key Controversies in the Philosophy of Science*. Chicago: University of Chicago Press.

———. 1990b. "Demystifying underdetermination." In *Scientific Theories*. Minnesota Studies in Philosophy of Science, vol. 14, ed. C. W. Savage, 267–97. Minneapolis: University of Minnesota Press.

———. 1996. *Beyond Positivism and Relativism: Theory, Method, and Evidence*. New York: Westview Press.

Lennox, J. G. 1992. "Teleology." In *Keywords in Evolutionary Biology*, eds. E. Fox-Keller and E. A. Lloyd, 324–33. Cambridge, Mass.: Harvard University Press.

Lenoir, T. 1989. *The Strategy of Life: Teleology and Mechanics in Nineteenth Century German Biology*. Chicago: University of Chicago Press. (Orig. publ. in Dordrecht, NL: D. Reidel Publishing, 1982.)

Leonardo da Vinci. 1987. *The Drawings and Miscellaneous Papers of Leonardo da Vinci in the Collection of Her Majesty the Queen at Windsor Castle*. New York: Harcourt Brace Jovanovich.

Leplin, J., ed. 1984. *Scientific Realism*. Berkeley: University of California Press.

———. 2000. *Realism and Instrumentalism: A Companion to the Philosophy of Science*. Ed. W. H. Newton-Smith, 393–401. Malden, Mass.: Blackwell Publishers.

Levitt, N. 1999. *Prometheus Bedeviled. Science and the Contradictions of Contemporary Culture*. New Brunswick: Rutgers University Press.

Lewontin, R. C. 1991. *Biology as Ideology: The Doctrine of DNA*. New York: Harper Collins.

Lindley, R. 1986. *Autonomy*. Atlantic Highlands, N.J.: Humanities Press International.

Longino, H. E. 1990. *Science as Social Knowledge*. Princeton: Princeton University Press.

———. 2001. *The Fate of Knowledge*. Princeton: Princeton University Press.

Löwy, I. 1997. "Participant observation and the study of biomedical sciences: Some methodological observations." In Söderqvist 1997, 91–107.

Lynch, M. 1985. *Art and Artifact in Laboratory Science: A Study of Shop Work and Shop Talk in a Research Laboratory*. London: Routledge and Kegan Paul.

———. 1993. *Scientific Practice and Ordinary Action: Ethnomethodology and Social Studies of Science*. Cambridge: Cambridge University Press.

Lynch, M. P. 1998. *Truth in Context: An Essay on Pluralism and Objectivity*. Cambridge, Mass.: MIT Press.

———, ed. 2001. *The Nature of Truth: Classic and Contemporary Perspectives*. Cambridge, Mass.: MIT Press.

Marcum, J. A. 2003. "Exploring the rational boundaries between the natural sciences and Christian theology." *Theology and Science* 1:203–20.

———. 2005a. "Metaphysical foundations and complementation of the natural sciences and theology." *Journal of Interdisciplinary Studies* 17:45–64.

———. 2005b. *Thomas Kuhn's Revolution: An Historical Philosophy of Science*. London: Continuum.

Marcuse, H. 1985. "On science and phenomenology." In *A Portrait of Twenty-five Years: Boston Colloquium for the Philosophy of Science 1960–1985*, eds. R. S. Cohen and M. W. Wartofsky, 19–30. Dordrecht, NL: D. Reidel Publishing.

Marx, L. 1964. *The Machine in the Garden: Technology and the Pastoral Ideal in America*. New York: Oxford University Press.

———. 1979. "Reflections on the neo-romantic critique of science." In *Limits of Scientific Inquiry,* eds. G. Holton and R. S. Morison, 61–74. New York: W. W. Norton.

Maslow, A. H. 1966. *The Psychology of Science*. New York: Harper and Row,

Masterman, M. 1970. "The nature of a paradigm." In *Criticism and Growth of Knowledge,* eds, I. Lakatos and A. Musgrave, 59–90. Cambridge: Cambridge University Press.

Matthews, G., M. Zeidner, and R. D. Roberts, eds. 2002. *Emotional Intelligence: Science and Myth*. Cambridge, Mass.: MIT Press.

Mayr, E. 1982. *The Growth of Biological Thought. Diversity. Evolution, and Inheritance*. Cambridge, Mass.: Harvard University Press.

———. 1988. "The multiple meanings of teleology." In *Towards a New Philosophy of Biology*, 38–66. Cambridge, Mass.: Harvard University Press.

———. 1992. "The idea of teleology." *Journal of the History of Ideas* 83:117–35.

McDowell, J. 1994. *Mind and World*. Cambridge, Mass.: Harvard University Press.

McFarland, T. 1969. *Coleridge and the Pantheist Tradition.* Oxford: Clarendon Press.

MacIntyre, A. 1978. "Objectivity in morals and objectivity in science." In *Morals, Science, and Sociality*, eds. H. T. Engelhardt Jr. and D. Callahan, 21–39. Hastings-on-Hudson, N.Y.: Hastings Center, Institute of Society, Ethics, and the Life Sciences.

———. 1984. *After Virtue: A Study in Moral Theory.* 2nd ed. South Bend, Ind.: University of Notre Dame Press.

McIntyre, L. C. 1993. "Complexity and social scientific laws." *Synthese* 97:209–27.

———. 2006. *Dark Ages: The Case for a Science of Human Behavior.* Cambridge, Mass.: MIT Press.

McMullin, E. 1983. "Values in science." *PSA* 2:3–28.

———. 1988. *Construction and Constraint: The Shaping of Scientific Rationality.* South Bend, Ind.: University of Notre Dame Press.

Megill, A., ed. 1994. *Rethinking Objectivity.* Durham, N.C.: Duke University Press.

Merton, R. K. 1973. *The Sociology of Science: Theoretical and Empirical Investigations.* Ed. N. W. Storer. Chicago: University of Chicago Press.

Merz, J. T. 1896/1965. *A History of European Thought in the Nineteenth Century*, vols. 1–4. New York: Dover.

Mikkelson, G. M. 2001. "Complexity and verisimilitude: Realism for ecology." *Biology and Philosophy* 16:533–46.

Milder, R. 1995. *Reimagining Thoreau.* Cambridge: Cambridge University Press.

Miller, C. 2002. "Realism, antirealism, and commonsense." In *Realism and Antirealism*, ed. W. P. Alston, 13–25. Ithaca: Cornell University Press.

Miller, D., trans. and ed. 1988. *Goethe: The Collected Works.* Vol. 12, *Scientific Studies.* Princeton: Princeton University Press.

Miller, K. R. 2004. "The flagellum unspun: The collapse of 'irreducible complexity.'" In Dembski and Ruse 2004, 81–97.

Miller-Rushing, A. J., and R. B. Primack. "Global warming and flowering times in Thoreau's Concord: a community perspective." *Ecology* 89:332–41.

Mishan, E. J. 1967. *The Costs of Economic Growth.* London: Staples Press.

Monod, J. 1971. *Chance and Necessity.* New York: Vintage.

Mooney, C. 2004. "Beware 'Sound Science.' It's Doublespeak for Trouble." *Washington Post*, February 29, page B02.

———. 2005. *The Republican War on Science.* New York: Basic Books.

Morrissey, R. J. 1988. "Introduction: Jean Starobinski and otherness." In *Jean-Jacques Rousseau: Transparency and Obstruction*, trans. A. Goldhammer. Chicago: University of Chicago Press, xiii–xxxviii.

Moser, P. K., ed. 1990. *Reality in Focus.* Englewood Cliffs, N.J.: Prentice-Hall.

Mosse, G. L. 1964. *The Crisis of German Ideology: Intellectual Origins of the Third Reich.* New York: Grosset and Dunlap.

Mueller, B. trans. and ed. 1989. *Goethe's Botanical Writings.* Woodbridge: Oxbow Press. Repr. University of Hawaii, 1952.

Mulhall, S. 2003. *Inheritance and Originality: Wittgenstein, Heidegger, Kierkegaard.* New York: Oxford University Press.

Nagel, T. 1986. *The View from Nowhere.* New York: Oxford University Press.

Nakhnikian, G. 2004. "It ain't necessarily so: An essay review of *Intelligent Design Creationism and its Critics: Philosophical, Theological, and Scientific Perspectives,*" by R. T. Pennock. *Philosophy of Science* 71:593–604.

Neiman, S. 1994. *The Unity of Reason: Rereading Kant.* New York: Oxford University Press.

Nelson, L. H. 1990. *Who Knows? From Quine to a Feminist Empiricism.* Philadelphia: Temple University Press.

Neuhouser, F. 1990. *Fichte's Theory of Subjectivity.* Cambridge: Cambridge University Press.

Neurath, O. 1931a/1959. "Sociology and 'physicalism.'" In A. J. Ayer 1959, 282–317.

———. 1931b/1983. "Physicalism." In R. Cohen and M. Neurath 1983, 52–57.

———. 1931c/1983. "Protocol statements." In R. Cohen and M. Neurath 1983, 91–99.

Nietzsche, F. 1967. *The Will to Power.* Trans. W. Kaufmann and R. J. Hollingdale. New York: Vintage Books.

Nitecki, M. H., and D. V. Nitecki. 1993. *Evolutionary Ethics.* Albany: State University of New York Press.

Nussbaum, M. C. 2001. *Upheavals of Thought: The Emotions.* Cambridge: Cambridge University Press.

Olson, R. 1991. *Science Deified and Science Defied: The Historical Significance of Science in Western Culture.* Vol. 2. of *From the Early Modern Age through the Early Romantic Era.* Berkeley: University of California Press.

Ortony, A., ed. 1993. *Metaphor and Thought.* Cambridge: Cambridge University Press.

Oyama, S. 1985. *The Ontogeny of Information.* Cambridge: Cambridge University Press.

Panksepp, J. 1998. *Affective Neuroscience: The Foundations of Human and Animal Emotions.* Oxford: Oxford University Press.

Peck, H. D. 1990. *Thoreau's Morning Work. Memory and Perception in A Week on the Concord and Merrimack Rivers, the Journal, and Walden.* New Haven: Yale University Press.

Pennock, R. T. 2001. *Intelligent Design Creationism and its Critics: Philosophical, Theological, and Scientific Perspectives.* Cambridge, Mass.: MIT Press.

Pick, D. 1997. "Stories of the eye." In *Rewriting the Self. Histories from the Renaissance to the Present,* ed. R. Porter, 186–99. London and New York: Routledge.

Pickering, A. 1994. "Objectivity and the mangle of practice." In Megill 1994, 109–24.

———, ed. 1992. *Science as Practice and Culture.* Chicago: University of Chicago Press.

———. 1995. *The Mangle of Practice.* Chicago: University of Chicago Press.

Pillow, K. 2000. *Sublime Understanding: Aesthetic Reflection in Kant and Hegel.* Cambridge, Mass.: MIT Press.

Pinch, T. 1986. *Confronting Nature: The Sociology of Neutrino Detection.* Dordrecht, NL: D. Reidel Publishing.

Pinker, S. 1997. *How the Mind Works.* New York: W. W. Norton.

Podolsky, S. H., and A. I. Tauber. 1997. *The Generation of Diversity, Clonal Selection Theory and the Rise of Molecular Immunology.* Cambridge, Mass.: Harvard University Press.

Polanyi, M. 1962. *Personal Knowledge. Towards a Post-critical Philosophy.* Corrected edition. Chicago: University of Chicago Press.

———. 1966. *The Tacit Dimension.* Garden City, N.Y.: Doubleday

Polkinghorne, J. C. 1985. *The Quantum World.* Princeton: Princeton University Press.

Poovey, M. 1998. *A History of the Modern Fact: Problems of Knowledge in the Sciences of Wealth and Society.* Chicago: University of Chicago Press.

Popper, K. R. 1935/1959. *The Logic of Scientific Discovery.* New York: Harper Torchbooks.

———. 1945. *The Open Society and its Enemies.* London: Routledge and Kegan Paul.

———. 1963. *Conjectures and Refutations: The Growth of Scientific Knowledge.* New York: Harper Torchbooks.

Postlethwaite, D. 1987. *Making it Whole: A Victorian Circle and the Shape of Their World.* Columbus: Ohio State University Press.

Prado, C. G., ed. 2003. *A House Divided. Comparing Analytic and Continental Philosophy.* Amherst, N.Y.: Humanity Books.

Price, J. J. 2001. *Thucydides and Internal War.* Cambridge: Cambridge University Press.

Proctor, R. N. 1988. *Racial Hygiene: Medicine Under the Nazis.* Cambridge, Mass.: Harvard University Press.

———. 1991. *Value-free Science? Purity and Power in Modern Knowledge.* Cambridge, Mass.: Harvard University Press.

Putnam, H. 1981. *Reason, Truth and History.* Cambridge: Cambridge University Press.

———. 1982. "Beyond the fact/value dichotomy." *Critica*, 3–12. Repr. in *Realism with a Human Face*, ed. J. Conant, 135–41. Cambridge, Mass., Harvard University Press, 1990.

———. 1983. "Why there isn't a ready-made world?" in *Realism and Reason: Philosophical Papers* 3:205–28. Cambridge: Cambridge University Press.

———. 1990. "Why is a philosopher?" In *Realism with a Human Face*, ed. J. Conant, 105–19. Cambridge, Mass.: Harvard University Press.

———. 1994. *Words and Life.* Ed. J. Conant. Cambridge, Mass.: Harvard University Press.

———. 2002. *The Collapse of the Fact/Value Dichotomy and Other Essays.* Cambridge, Mass.: Harvard University Press.

———. 2008. *Jewish Philosophy as a Way of Life. Rosenzweig, Buber, Levinas, Wittgenstein*. Bloomington: Indiana University Press.

———. 2009. "A Proof on the 'Undetermination Doctrine.'" In *Hilary Putnam. Philosophy in an Age of Science: Physics, Mathematics, and Skepticism*, eds. M. De Caro and D. MacArthur. Cambridge, Harvard University Press, forthcoming.

Putnam, R. A. 1985. "Creating facts and values." *Philosophy* 60:187–204.

Quine, W. V. O. 1953a/1961/1980. "Two dogmas of empiricism." In *From a Logical Point of View*. 2nd ed., 20–46. Cambridge, Mass.: Harvard University Press.

———. 1953b/1961/1980. Foreword to 1980. In *From a Logical Point of View*. 2nd ed. Cambridge, Mass.: Harvard University Press.

———. 1953c/1961/1980. "Identity, ostension, and hypostasis." In *From a Logical Point of View*. 2nd ed., 65–79. Cambridge, Mass.: Harvard University Press.

———. 1953d/1976. "Mr. Strawson on logical theory." In *The Ways of Paradox and other Essays*, 137–57. Cambridge, Mass.: Harvard University Press.

———. 1969a. "Natural kinds." In *Ontological Relativity and Other Essays*, 114–38. New York: Columbia University Press.

———. 1969b. "Ontological relativity." In *Ontological Relativity and Other Essays*, 26–68. New York: Columbia University Press.

———. 1969c. "Epistemology naturalized." In *Ontological Relativity and Other Essays*, 69–90. New York: Columbia University Press.

———. 1975. "On empirically equivalent systems of the world." *Erkenntnis* 9:313–28.

———. 1990. "Three indeterminancies." In *Perspectives on Quine*, eds. R. B. Barrett and R. F. Gibson, 1–16. London: Blackwell.

———. 1991. "Two dogmas in retrospect." *Canadian Journal of Philosophy* 21:265–74.

———. 1995. *From Stimulus to Science*. Cambridge, Mass.: Harvard University Press.

Quine, W. V. O., and J. S. Ullian. 1978. *The Web of Belief*. 2nd ed. New York: Random House.

Railton, P. 2003. *Facts, Values, and Norms: Essays towards a Morality of Consequence*. Cambridge: Cambridge University Press.

Rehbock, R. F. 1983. *The Philosophical Naturalists: Themes in Early Nineteenth Century British Biology*. Madison: University of Wisconsin Press.

Reichenbach, H. 1938. *Experience and Prediction: An Analysis of the Foundations and the Structure of Knowledge*. Chicago: University of Chicago Press.

———. 1951. *The Rise of Scientific Philosophy*. Berkeley: University of California Press.

Rentschler, L., B. Herzberger, and D. Epstein, eds. 1988. *Beauty and the Brain: Biological Aspects of Aesthetics*. Basel, CH: Birkhauser Verlag.

Rescher, N. 1987. *Scientific Realism*. Dordrecht, NL: D. Reidel Publishing.

Richards, R. J. 1987. "Appendix I." In *Darwin and the Emergence of Evolutionary Theories of Mind and Behavior*, 559–91. Chicago: University of Chicago Press.

Ridley, M. 1997. *The Origins of Virtue: Human Instincts and the Evolution of Cooperation*. New York: Viking Books.

Rockmore, T. 2005. *Hegel, Idealism, and Analytical Philosophy*. New Haven: Yale University Press.

Rolston, H. 1988. *Environmental Ethics: Duties to and Values in the Natural World*. Philadelphia: Temple University Press.

Romanos, G. D. 1983. *Quine and Analytic Philosophy: The Language of Language*. Cambridge, Mass.: MIT Press.

Root-Bernstein, R. 1989. *Discovering: Inventing and Solving Problems at the Frontiers of Scientific Knowledge*. Cambrdge, Mass.: Harvard University Press.

Rorty, R. 1991a. "Solidarity or objectivity." In *Objectivity, Relativism, and Truth. Philosophical Papers*, vol. 1, 21–34. Cambridge: Cambridge University Press.

———. 1991b. "Science as solidarity." In *Objectivity, Relativism, and Truth. Philosophical Papers*, vol. 1, 35–45. Cambridge: Cambridge University Press.

———. 2002. "Worlds or words apart? The consequences of pragmatism for literary studies: An interview with Richard Rorty (with E. P. Ragg)." *Philosophy and Literature* 26:369–96.

Rosenberg, A. 1985. *The Structure of Biological Science*. Cambridge: Cambridge University Press.

Ross, A. 1996. *Science Wars*. Durham, N.C.: Duke University Press.

Roszak, T. 1972. *Where the Wasteland Ends*. Garden City, N.Y.: Doubleday.

Rouse, J. 1987. *Knowledge and Power: Toward a Political Philosophy of Science*. Ithaca: Cornell University Press.

Rudd, A. 2003. *Expressing the World: Skepticism, Wittgenstein, and Heidegger*. Peru, Ill.: Open Court.

Rudwick, M. 1985. *The Great Devonian Controversy: The Shaping of Scientific Knowledge among Gentlemanly Specialists*. Chicago: University of Chicago Press.

Ruse, M. 1993. "The significance of evolution." In *A Companion to Ethics*, ed. P. Singer, 500–510. Oxford: Blackwell.

———. 1996. *Monad to Man: The Concept of Progress in Evolutionary Biology*. Cambridge, Mass.: Harvard University Press.

———. 1999. *Mystery of Mysteries: Is Evolution a Social Construction?* Cambridge, Mass.: Harvard University Press.

———. 2003. *Darwin and Design: Does Evolution have a Purpose?* Cambridge, Mass.: Harvard University Press.

Russell, B. 1935/1978. *Science and Religion*. Oxford: Oxford University Press.

Sadler, J. Z. 1997. "Recognizing values: A descriptive-causal method for medical/scientific discourses." *Journal of Medicine and Philosophy* 22:541–65.

Sarkar, S, and A. I. Tauber. 1991. "Fallacious claims for HGP." *Nature* 353:691.

Sartre, J-P. 1943/1956. *Being and Nothingness: An Essay on Phenomenological Ontology*. Trans. H. E. Barnes. New York: Philosophical Library.

Sayre-McCord, G. 1988. "Introduction: The many moral realisms." In *Essays on Moral Realism*, ed. G. Sayre-McCord, 1–23. Ithaca: Cornell University Press.

Schaffner, K. F. 1993. *Discovery and Explanation in Biology and Medicine*. Chicago: University of Chicago Press.

———. 2002. "Assessments of efficacy in biomedicine. The turn toward methodological pluralism." In *The Role of Complementary and Alternative Medicine: Accommodating Pluralism*, ed. D. Callahan, 1–14. Washington, D.C.: Georgetown University Press.

Schatzki, T. R., K. Knorr Cetina, and E. von Savigny, eds. 2001. *The Practice Turn in Contemporary Theory*. London: Routledge.

Schiller, F. 1801/1993. "Letters on the Aesthetic Education of Man." In *Essays*, trans. E. M. Wilkinson and L. A. Willoughby, eds. W. Hinderer and D. O. Dahlstrom, 86–178. New York: Continuum.

Schönborn, C. 2005. "Finding design in nature." *The New York Times*, July 7.

Schopenhauer, A. 1851/1974. *Parerga and Parilipomena, Short Philosophical Essays*, vols. 1 and 2. Trans. E. F. J. Payne. Oxford: Oxford University Press.

Schrödinger, E. 1944/1967. *What is Life?* Cambridge: Cambridge University Press.

Schwartz, R. S. 1995. Review of *The Golden Helix: Inside Biotech Ventures* by A. Kornberg, *New England Journal of Medicine* 333:1292–93.

Segerstrale, U. 2000a. *Defenders of the Truth: The Sociobiology Debate*. New York: Oxford University Press.

———. 2000b. *Beyond the Science Wars: The Missing Discourse about Science and Society*. Albany: State University of New York Press.

Sellars, W. 1963. *Science, Perception, and Reality*. New York: Humanities Press.

———. 1997. *Empiricism and the Philosophy of Mind*. Cambridge, Mass.: Harvard University Press.

Shapin, S. 1982. "History of science and its sociological reconstructions." *History of Science* 20:157–211.

———. 1994. *A Social History of Truth: Civility and Science in Seventeenth-Century England*. Chicago: University of Chicago Press.

———. 1996. *The Scientific Revolution*. Chicago: University of Chicago Press.

———. 2001. "How to be antiscientific." In Labinger and Collins 2001, 99–115.

Shapin, S., and S. Schaffer. 1985. *Leviathan and the Air Pump: Hobbes, Boyle and the Experimental Life*. Princeton: Princeton University Press.

Shattuck, R. 1996. *Forbidden Knowledge: From Prometheus to Pornography*. New York: Harvest Books.

Shimony, A. 1978, 1991. "Metaphysical problems in the foundations of quantum mechanics." *International Philosophical Quarterly* 8:2-17. Reprinted in *The Philosophy of Science*, eds. R. Boyd, P. Gasper, and J. D. Trout, 517–28. Cambridge: The MIT Press.

Shook, J. R. 2000. *Dewey's Theory of Knowledge and Empirical Reality*. Nashville: Vanderbilt University Press.

Simon, H. 1969. *The Science of the Artificial*. Cambridge, Mass.: MIT Press.

Simon, W. M. 1963. *European Positivism in the Nineteenth Century*. Ithaca: Cornell University Press.

Smith, R. 1997. *The Norton History of the Human Sciences.* New York: W. W. Norton.

———. 2007. *Being Human: Historical Knowledge and the Creation of Human Nature.* Manchester: Manchester University Press.

Snow, C. P. 1959/1964. *The Two Cultures and a Second Look.* Cambridge: Cambridge University Press.

Soames, S. 2003. *Philosophical Analysis in the Twentieth Century.* Vol. 2 of *The Age of Meaning.* Princeton: Princeton University Press.

Sober, E. 1993. *Philosophy of Biology.* Boulder: Westview Press.

Söderqvist, T. 2003. *Science as Autobiography: The Troubled Life of Niels Jerne.* New Haven: Yale University Press.

———, ed. 1997. *The Historiography of Contemporary Science and Technology.* Amsterdam: Harwood Academic Publishers.

Sokal, A. 1996. "Transgressing the boundaries: Towards a transformative hermeneutics of quantum gravity." *Social Text* 46/47:217–52.

———. 2008. *Beyond the Hoax: Science, Philosophy, and Culture.* New York: Oxford University Press.

Sorell, T. 1991. *Scientism: Philosophy and the Infatuation with Science.* London: Routledge.

Stempsey, W. E. 1999. *Disease and Diagnosis: Value-dependent Realism.* Dordrecht, NL: Kluwer Academic Publishers.

Tait, W. W., ed. 1997. *Early Analytic Philosophy: Frege, Russell, Wittgenstein.* Chicago: Open Court.

Tauber, A. I. 1993. "Goethe's philosophy of science: Modern resonances." *Perspectives in Biology and Medicine* 36:244–57.

———. 1994. *The Immune Self: Theory or Metaphor?* New York: Cambridge University Press.

———. 1995. "Postmodernism and immune selfhood." *Science in Context* 8:579–607.

———. 1996a. "From Descartes' dream to Husserl's nightmare." In Tauber 1996b, 289–312.

———, ed. 1996b. *The Elusive Synthesis: Aesthetics and Science.* Dordrecht, NL: Kluwer Academic Publishers.

———. 1998. "Ecology and the claims for a science-based ethics." In *Philosophies of Nature: The Human Dimension in Celebration of Erazim Kohák.* Boston Studies in the Philosophy of Science 195, eds. R. S. Cohen and A. I. Tauber, 185–206. Dordrecht, NL: Kluwer Academic Publishers.

———. 1999a. *Confessions of a Medicine Man: An Essay in Popular Philosophy.* Cambridge, Mass.: MIT Press.

———. 1999b. Review of *The Historiography of Contemporary Science and Technology*, ed. T. Söderqvist. *Science, Technology, and Human Values* 24:384–401.

———. 1999c. "Is biology a political science?" *BioScience* 49:479–86.

———. 2000. Review of *Prometheus Bedeviled: Science and the Contradictions of Contemporary Society* by Norman Levitt. *Science, Technology, and Human Values* 25:385–89.

———. 2001. *Henry David Thoreau and the Moral Agency of Knowing*. Berkeley: University of California Press.

———. 2003. "The philosopher as prophet: The case of Emerson and Thoreau." *Philosophy in the Contemporary World* 10:89–103.

———. 2005a. *Patient Autonomy and the Ethics of Responsibility*. Cambridge, Mass.: MIT Press.

———. 2005b. "The biological notion of self and nonself." *Stanford Encyclopedia of Science*. Ed. E. N. Zalta. http://plato.stanford.edu/entries/biology-self/.

———. 2007. "Science and reason, reason and faith: A Kantian perspective." In *Intelligent Design: Science or Religion? Critical Perspectives*, eds. R. M. Baird and S. E. Rosenbaum, 307–36. Amherst, N.Y.: Prometheus Books.

Tauber, A. I., and L. Chernyak. 1991. *Metchnikoff and the Origins of Immunology*. New York: Oxford University Press.

Tauber, A. I., and S. Sarkar. 1992. "The Human Genome Project: Has blind reductionism gone too far?" *Perspectives in Biology and Medicine* 35:220–35.

———. 1993. "The ideological basis of the Human Genome Project." *Journal of the Royal Society of Medicine* 86:537–40.

Taylor, C. 1970. "The explanation of purposive behavior." In *Explanation in the Behavioural Sciences: Confrontations*, eds. R. Borger and F. Cioffi, 49–79. Cambridge: Cambridge University Press.

Taylor, P. W. 1986. *Respect for Nature: A Theory of Environmental Ethics*. Princeton: Princeton University Press.

Thagard, P. 2000, *Coherence in Thought and Action*. Cambridge, Mass.: MIT Press.

———. 2002. "The passionate scientist: Emotion in scientific cognition." In Carruthers, Stich, and Siegal 2002, 235–50.

Thoreau, H. D. 1962. *The Journal of Henry David Thoreau*. Eds. B. Torrey and F. H. Allen. New York: Dover Books.

———. 1971. *The Writings of Henry David Thoreau: Walden*. Ed. J. L. Shanley. Princeton: Princeton University Press.

———. 1980. "Walking." In *The Natural History Essays*, 93–136. Salt Lake City: Gibbs-Smith Publisher.

———. 1981. *The Writings of Henry David Thoreau: Journal vol. 1, 1837–1844*. Eds. E. H. Witherell, W. L. Howarth, R. Sattelmeyer, and T. Blanding. Princeton: Princeton University Press.

———. 1984. *The Writings of Henry David Thoreau: Journal vol. 2, 1842–1848*. Ed. R. Sattelmeyer. Princeton: Princeton University Press.

———. 1990. *The Writings of Henry David Thoreau: Journal vol. 3, 1848–1851*. Eds. R. Sattelmeyer, M. R. Patterson, and W. Rossi. Princeton: Princeton University Press.

———. 1992. *The Writings of Henry David Thoreau: Journal vol. 4, 1851–1852*. Eds. L. N. Neufeldt and N. C. Simmons. Princeton: Princeton University Press.

———. 1993. "The dispersion of seeds." In *Faith in a Seed: The Dispersion of Seeds and Other Late Natural History Writings*, ed. B. P. Dean, 23–173. Washington, D.C.: Island Press.

———. 1997. *The Writings of Henry David Thoreau: Journal vol. 5, 1852–1853.* Ed. P. F. O'Connell. Princeton: Princeton University Press.

Toulmin, S. 2001. *Return to Reason.* Cambridge, Mass.: Harvard University Press.

Trigg, R. 1980. *Reality at Risk: A Defense of Realism in Philosophy and the Sciences.* Brighton: Harvester Press.

Trudeau, G. 2006. Doonesbury, *Slate*, March 5. http://www.doonesbury.com/strip/dailydose/index.html?uc_full_date=20060305.

Van Fraassen, B. 1980. *The Scientific Image.* Oxford: Clarendon Press.

Walls, L. D. 1995. *Seeing New Worlds: Henry David Thoreau and Nineteenth-century Natural Science.* Madison: University of Wisconsin Press.

Watson, J. D. 1968. *The Double Helix.* New York: Atheneum Press.

Weber, M. 1904/1949. "'Objectivity' in social science and social policy." In *The Methodology of the Social Sciences*, trans. and eds. E. A. Shils and H. A. Finch, 50–112. New York: Free Press.

———. 1919/1946. "Science as a vocation." In *From Max Weber: Essays in Sociology*, trans. and eds. H. H. Gerth and C. W. Mills, 137–56. New York: Oxford University Press.

Weinberg, S. 1992. *Dreams of a Final Theory.* New York: Pantheon.

Wendling, K. 1996. "Is science unique?" *Biology and Philosophy* 11:421–38.

Whewell, W. 1840. *Philosophy of the Inductive Sciences.* London: J. W. Parker.

Whitehead, A. 1925. *Science and the Modern World.* London: Macmillan.

Whitman, W. 1865/1902. "By the Roadside." In *The Complete Writings of Walt Whitman,* 25–39. New York: G. P. Putnam's Sons.

Wilson, A. N. 1999. *God's Funeral: A Biography of Faith and Doubt in Western Civilization.* New York: Ballantine Books.

Wilson, E. O. 1978. *Human Nature.* Cambridge, Mass.: Harvard University Press.

———. 1998. *Consilience.* New York: Vintage Books.

Wittgenstein, L. 1921/1961. *Tractatus Logico-Philosophicus.* Trans. D. F. Pears and B. F. McGuiness. London: Routledge.

———. 1953/1968. *Philosophical Investigations.* 3rd ed. Trans. G. E. M. Anscombe. New York: Macmillan.

———. 1969. *On Certainty.* Eds. G. E. M. Anscombe and G. H. von Wright. New York: Harper Row.

Wolpert, L., and A. Richards. 1997. *Passionate Minds: The Inner World of Scientists.* Oxford: Oxford University Press.

Wolterstorff, N. 1984. "Realism vs. anti-realism." In *Realism Proceedings and Addresses of the American Catholic Philosophical Association*, vol. 59, ed. D. O. Dahlstrom, 182–205. Washington, D.C.: The American Catholic Philosophical Association.

Woolgar, S. 1988a. *Science: The Very Idea.* Chichester, UK: Ellis Harwood.

———, ed. 1988b. *Knowledge and Reflexivity: New Frontiers in the Sociology of Knowledge.* London: Sage Publications.

Wright, L. 1973. "Functions." *Philosophical Review* 82:139–68.

———. 1976. *Ideological Explanations.* Berkeley: University of California Press.

Zammito, J. H. 2004. *A Nice Derangement of Epistemes: Post-positivism in the Study of Science from Quine to Latour.* Chicago: University of Chicago Press.

Index